'Understanding statistics means being ⌐
phenomena that surround us, and tl
statistical concepts and ending in cor
excellent primer on essential statistical knowledge. Thanks to its clarity and practical
examples included, this book is easy to read and understand, which makes it
appealing for various readers. This book approaches main statistical concepts in an
accessible and understandable way and is an extremely valuable reading for students,
professionals, and researchers. I am sure that a wide audience will find this book as
"a must have" in their regular work.'
Dr Iva Sverko, *Institute of Social Sciences Ivo Pilar, Zagreb, Croatia*

Interpreting Statistics for Beginners

Interpreting Statistics for Beginners teaches readers to correctly read and interpret results of basic statistical procedures as they are presented in scientific literature, and to understand what they can and cannot infer from such results.

The first of its kind, this book explains key elements of scientific paradigms and philosophical concepts that the use of statistics is based on and introduces readers to basic statistical concepts, descriptive statistics and basic elements and procedures of inferential statistics. Explanations are accompanied with detailed examples from scientific publications to demonstrate how the procedures are used and correctly interpreted. Additionally, *Interpreting Statistics for Beginners* shows readers how to recognize pseudoscientific claims that use statistics or statements not based on the presented data, which is an important skill for every professional relying on statistics in their work.

Written in an easy-to-read style and focusing on explaining concepts behind statistical calculations, the book is most helpful for readers with no previous training in statistics, and also those wishing to bridge the conceptual gap between doing the statistical calculations and interpreting the results.

Prof. Dr. Vladimir Hedrih is a full professor of psychology at the University of Niš in Serbia. He is also the author of the book *Adapting Psychological Tests and Measurement Instruments for Cross-Cultural Research: An Introduction.*

Dr. Andjelka Hedrih is a research assistant professor at the Mathematical Institute of the Serbian Academy of Sciences and Arts. She is a medical doctor who holds a PhD in Biomedical engineering and technologies. She is the author of a number of published papers on mathematical modelling of biological systems.

Interpreting Statistics for Beginners

A Guide for Behavioural and Social Scientists

Vladimir Hedrih and Andjelka Hedrih

Routledge
Taylor & Francis Group

LONDON AND NEW YORK

Cover image: Getty Images

First published 2022
by Routledge
4 Park Square, Milton Park, Abingdon, Oxon OX14 4RN

and by Routledge
605 Third Avenue, New York, NY 10158

Routledge is an imprint of the Taylor & Francis Group, an informa business

British Library Cataloguing-in-Publication Data
A catalogue record for this book is available from the British Library

Library of Congress Cataloguing-in-Publication Data
Names: Hedrih, Vladimir, author. | Hedrih, Andjelka, 1978- author.
Title: Interpreting statistics for beginners: a guide for behavioural and social scientists / Vladimir Hedrih, Andjelka Hedrih.
Description: Abingdon, Oxon; New York, NY: Routledge, 2022. | Includes bibliographical references and index. |
Identifiers: LCCN 2021044964 (print) | LCCN 2021044965 (ebook) | ISBN 9780367618520 (paperback) | ISBN 9780367620516 (hardback) | ISBN 9781003107712 (ebook)
Subjects: LCSH: Social sciences--Statistical methods. | Psychology--Statistical methods. | AMS: Statistics. | Statistics -- General reference works (handbooks, dictionaries, bibliographies, etc.). | Statistics -- Instructional exposition (textbooks, tutorial papers, etc.).
Classification: LCC HA29.H426 2022 (print) | LCC HA29 (ebook) | DDC 300.72/7--dc23
LC record available at https://lccn.loc.gov/2021044964
LC ebook record available at https://lccn.loc.gov/2021044965

ISBN: 978-0-367-62051-6 (hbk)
ISBN: 978-0-367-61852-0 (pbk)
ISBN: 978-1-003-10771-2 (ebk)

DOI: 10.4324/9781003107712

Typeset in Bembo
by MPS Limited, Dehradun

Contents

Preface

Dear readers,

We live in the age of science and most people in their work will more or less often have a need to read and interpret statistical data and results of statistical calculations. Almost as often there will be a need to verify the validity of conclusions and statements others have derived from statistical data. The ability to derive correct conclusions in such situations can be the thing making a difference between success and failure in the task the person is working on. Most of the people, most of the time, are users of statistical calculations and results others have produced, while the number of situations where people are doing calculations themselves is much smaller. However, most of the introductory statistical literature is written with the goal of teaching the reader how to do statistical calculations him/herself while devoting much less attention to the interpretation of results.

Recognizing this situation, we set out to devise a plan for teaching statistics that would be primarily focused on what we called statistical literacy, that is, teaching the valid ways of interpreting and understanding the results of statistical calculations, teaching what conclusions can and cannot be derived from what types of data and not only teaching readers and students to recognize when conclusions drawn from a set of data are valid but also when they are invalid or baseless or insufficiently substantiated. This book is the result of this plan. On the one side, this is an introductory book in statistics, but, on the other, it is a book that, rather than on statistical calculations and formulae, focuses on presenting to the reader the basic concepts and ideas statistical calculations are based on and the general scientific and epistemological framework in which statistical calculations are done and used. It is written for people who are just starting to learn statistics, or people who are not into statistics at all but need to be able to understand and interpret statistical data in their work. Due to this, the book also features examples of statistical data and its interpretations from a wide range of topics in the area of behavioural and social sciences. This is also a book that covers the introductory statistics course that the first author teaches at his university and, in creating the concept for this book, more than 20 years of experience in developing this course have been condensed into these pages.

Enjoy!
The authors

Acknowledgements

Authors would like to thank Elsevier, the Society for Physical Therapy Science, Dr Tomas Sobotka, Dr Stanislava Popov, Dr Qasim Ayub, Dr Ana Radanović, Dr Urša Čizman Štaba, Dr Sonja Čotar Konrad, Dr Dan Zhang and Dr Maša Vukčević Marković for their permissions to use tables and graphs from their published papers in this book. We would also like to thank Ms Ana Jovančević and Dr Ivana Pedović for their extensive help with choosing the artwork for the book covers.

Before we start: General instructions for the exercises

The goal of this book is to help the reader learn how to read and interpret the results of statistical calculations presented in scientific literature. Due to this, at the end of every chapter in the book (except the first one), there are exercises that provide the reader with an opportunity to apply the statistical knowledge that has been covered by the book so far. There is also the last chapter consisting exclusively of such exercises. These exercises consist of excerpts from real scientific publications and of stories describing research plans. We will jointly be referring to these as excerpts. Each excerpt is accompanied by a number of statements referring to its contents that the reader is asked to categorize into different categories based on its properties and how it relates to the contents of the excerpt.

Please examine each excerpt and read each statement carefully and categorize it into one of four categories:

1   **True** – if it can be seen or concluded from the excerpt that the statement is **true**.
2   **False** – if the statement **is meaningful (**see the description of meaningless statements), but **it can be concluded from the excerpt that it is false**.
3   **Meaningless** – if the statement is **meaningless**, i.e. if:

   - modern statistical procedures and scientific methods cannot, not even theoretically, produce a result that could confirm it and/or
   - if the statement is logically impossible or grammatically meaningless and/or
   - if the statement mentions some non-existent or meaningless statistical term or concept
   - if it states that some existing/meaningful statistic has an impossible value (a value it cannot have)

4   **Unknown/cannot be inferred from the data available** – if the statement is meaningful i.e. such that existing statistical and scientific procedures can (at least theoretically) produce a results that could confirm it, but there is no data on the topic in the excerpt and therefore it cannot be concluded based on the excerpt whether the statement is true or not.

The following rules will also apply to these exercises:

- Statements will not contain meaningful-existing statistical terms that are not explained in this book.
- Unless the opposite is apparent from the data in the excerpt, it should be assumed that all the conditions for the use of the procedure presented in the excerpt have been met, i.e. that presented statistical procedures are applied correctly.
- Statistically significant, if not explicitly declared otherwise, means that .05 should be used as the threshold.
- Unless explicitly stated otherwise in the specific statements, when referring to variables in the excerpt (in a general way), the variable dividing the sample into groups will not be counted among them (if such a variable exists i.e. if the sample is divided into groups, even if its name is given in the excerpt).
- If not explicitly declared otherwise, it should be assumed that values of the variable were positively scored (with higher scores indicating higher levels of the construct the variable refers to).

This is the general instruction that is valid for all exercises of this type found throughout the book. After each exercise, there is a presentation of correct answers and a short explanation of each one. Read them to find out how well you performed and clarify any mistakes.

1 Scientific paradigms and scientific explanations, pseudoscience

In the modern world, we use statistics in a great multitude of areas. Every time that we need to describe a large set of data, people or cases or when we need to make inferences about topics that involve large sets of cases, we refer to statistics. By definition, statistics is a science, more precisely a branch of mathematics, that deals with collecting and analyzing large sets of numerical data. It allows us to describe such sets and make inferences from them. As a part of science, good practice of statistics relies on the adequate application of the scientific methods and also on understanding the conditions and limitations its use is based on. To understand how to use statistics adequately, the person must understand the assumptions it is based on and the nature of results it can produce. This is the reason why the first chapter of this book, the goal of which is to introduce the reader to the use of statistics, will be devoted to discussing the functioning of science and considering the place of statistics in it. In this chapter, we will also draw the line and define the boundary between good scientific practice and malpractices that try to use the good name of science but are not scientific at all.

1.1 Science, scientific paradigms and statistics

Science can be defined as a systematic enterprise that builds and organizes knowledge in the form of testable explanations and predictions about the universe (*Science* (80-.)., 2020). As its main result, science produces scientific generalizations, laws and theories that can be tested. Ideally, these products of science are based on the interpretation of relations between scientific observations i.e. empirical facts collected in the process of scientific research. However, what constitutes a scientific observation and what does not constitute one i.e. what can be considered an empirical fact and what cannot and also where a scientist will look for scientific observations and what types of scientific observations, is something defined, often implicitly, by the scientific paradigm the scientists are using. A scientific **paradigm** is a philosophical and theoretical frame of a certain school of scientific thought or a scientific discipline that serves as a basis for formulated theories, laws and generalizations and also organizing scientific observations. The term was introduced into science during the 20th century (e.g. Walker, 2010; Wray, 2011) and was hugely popularized by Thomas Kuhn who defined it as a "universally recognized scientific achievement that for a time provides model problems and solutions to a community of practitioners" (Kuhn, 1970, p. viii). Typically, a paradigm is a set of beliefs that scientists in a certain area share that can most often not be verified at all, yet it both defines the types of observations scientists will be looking for and what can be inferred from them and how. These beliefs are often implicit and will often be taken for

DOI: 10.4324/9781003107712-1

granted and seen as the only valid worldview by those accepting them. Because paradigms are the basis of all scientific knowledge, although they themselves remain unverified and unverifiable, it is very important that scientists and people working with scientific data be aware of the assumptions of the paradigm or paradigms they are using so they can recognize the properties and limits of scientific knowledge they are dealing with. Kuhn believed, and most philosophers of science today would agree (Kuhn, 1970; Walker, 2010; Wray, 2011), that paradigms tend to change with time and that this tends to happen at times when it no longer becomes possible to ignore scientific observations that cannot be adequately explained using the existing paradigm and when, at the same time, there is another paradigm that can provide a better explanation of these observations available. The times when the science works within the paradigm were called periods of "Normal science" by Kuhn, while the brief periods when the change in paradigms happens were named "scientific revolutions" by Kuhn. It was his belief that history of science can be described by long periods of "normal science" interspersed by short and dynamic periods of "scientific revolutions". After a scientific revolution happens, a period of normal science would start again with science now working within the scope of the newly acquired paradigm.

What are the assumptions of the current scientific paradigm? Most scientific fields have their own sets of assumptions that form the specific paradigms used in their fields. This is especially true if we take Kuhn's definition of a paradigm as a set of model problems and solutions as these are indeed specific for individual scientific fields. This specificity of paradigms of individual fields of science is actually what led to Kuhn's discovery of paradigms – Kuhn states that he realized the importance of paradigms when he became aware of the large differences that existed between social scientists with regard to assumptions about the nature of the scientific problems they were dealing with (Kuhn, 1970). However, if we look at the situation more broadly, we can notice that aside from paradigmatic elements that are characteristic for every scientific field, there are also (unverified and unverifiable) assumptions that the whole science in general is based on and that are shared by various scientific fields. We will try to provide here a brief overview of these assumptions, especially of the ones that are the most relevant for understanding the science and practice of statistics that will be presented in this book, but the reader should be aware that, as paradigm elements are typically implicit, this should not be taken to represent an exhaustive list of all common properties of the current scientific paradigm. Here is the list:

- There is an **objective reality outside of our mind** and we can adequately perceive it using our recognized senses;
- A phenomenon is a part of the objective reality i.e. it exists, only if **it can be perceived by multiple persons**. If a phenomenon can be perceived by only a single person and by no one else, it is not real;
- Our mind is separate from the objective reality and does not influence it, except through actions of our recognized organs;
- Spacetime is homogenous – neither space nor time are factors in anything by themselves – all scientific laws work equally in all points of space and time. Laws of nature do not change with either space or time, they function equally everywhere;
- the functioning of the universe can be completely described by a set of rules that is sufficiently simple to fit inside the universe;

- Time is a special dimension – we can only move forward through it (from past to future), unlike the space dimensions through which we can move in both directions.
- Causality functions in one temporal direction only, does not function in the opposite direction – past events can influence future events, but present or future events cannot influence past events. Present and future cannot change the past;
- The phenomena we are observing have causes (that lead to their appearance) and it is, in principle, possible to determine what those causes are.
- And so on...

The objective reality outside of our minds. In its essence, scientific research is based on observation and observation is only meaningful if there is something to observe and if there is a way in which we can observe it. Therefore, a necessary underlying assumption for such an undertaking is that there is something outside of our mind that we can observe i.e. the objective reality, and also that our senses are adequate tools for observing i.e. discovering properties of that outside reality. However, as lots of works in the philosophy of science and also the coverage of the topic in the works of fiction teach us, we really have no indisputable evidence that what we are perceiving is an objective reality outside of our minds observed through the use of our senses and not something else that could also produce a stream of conscious experiences – such as a simulation, a fabrication of our minds etc. The situation is probably best described by the philosophical stance of solipsism and intense discussions it has been raising throughout history (e.g. Baldwin & Bell, 1988; Fodor, 1991). To make things more complicated, we do not consider all perceptions we experience to be observations of the outside reality. All of us have perceptions that we experience but that we do not take to be observations of the outside reality. Such perceptions include those in dreams, hallucinations, "visions", products of imagination, but also inaccurate observations. For the purposes of scientific research, it is important for a person to be able to differentiate between perceptions that are the result of observing the external reality and those that are not and that scientists be in agreement about the criteria for this differentiation. For this reason, the current convention in science is that only perceptions that have been obtained through the use of sensory organs of the human body, those organs that science recognizes and understands can be considered to constitute observations. These organs can be helped by using various instruments, but for a perception to be considered as a basis of scientific observation it must in the end be perceived by a recognized sensory organ. Those perceptions that were not obtained through recognized sensory organs are not considered observations and cannot be the bases of scientific work. When saying this, we emphasize the words "recognized sensory organs", as throughout history, there were differing opinions as to what methods of obtaining perceptions were considered valid. There were places and times where, for example, drug-induced hallucinatory experiences were considered sources of valid observations (e.g. Wallace, 1959). Also, as science progressed and learned more about how our sensorium functions, the number of recognized senses increased – while people have probably been aware of the function of eyes and ears science the dawn of history, some other sensory organs, such as those responsible for proprioception (perception of the motion and position of our body and its parts) or the function of the inner ear, have been discovered relatively recently. Due to this, the number of sensory organs science recognized has also increased with the advancements of science. The implicit assumption in science is that if an observation did not happen through a recognized sensory organ, it is not considered valid or "real". Only observations obtained through our (recognized) senses are eligible to be considered valid.

The intersubjective agreement principle. A separate issue is that of what happens when observations of the same phenomenon by two different persons differ. Or what happens when a phenomenon can be observed by one person, but not by another? The simplest way to differentiate valid observations from inaccurate observations, hallucinations and similar phenomena is that valid observations performed by different people will be similar. Phenomena that one person claims to have observed, but others have not been able to observe, although they were in a position to do so, are not real. The underlying assumption of science is that if something is "real" and if it exists in the physical world, it is something that can be observed by multiple persons. If it cannot be observed by more than a single person or a specific group of persons, we cannot be sure whether they are making false (made up) claims or whether they have had inaccurate observations, hallucinations or something similar, but the scientific assumption would be that there is something wrong with the observations. Only if a phenomenon can be observed by multiple persons, ideally, by anyone who is interested in observing it, can such observations and conclusions from such observations be considered scientifically valid. This is the principle of intersubjective agreement which is the fundamental component of both scientific objectivity and replicability of scientific findings, both key components of good science. However, as practically useful as it typically is, the assumption that only phenomena that are observable by multiple people are real also creates its own sets of issues for science. According to this principle, phenomena that would happen only once, i.e. **unique phenomena would not be scientifically observable**. They would, in other words, not be considered real. The same goes for phenomena that can be observed by only one person, because no other person is in a position to observe them. But do these types of phenomena really exist? While there can be a debate about whether some proposed unique, one-off events reported by various people are real or made up, there are ample situations in the history of science where phenomena that were observed and reported by just a single individual were considered hoaxes or made-up claims, only to be accepted as real by the scientific community after it became possible for other people to observe them. A recent example of this situation might be the phenomenon of ball lightning, a natural phenomenon that was considered a fairy-tale, a legend or a hoax by many (in spite of multiple historical reports), until it was captured on camera and using scientific instruments by a group of scientists (Cen et al., 2014). However, ball lightning happened to be a rare phenomenon, but not a unique or a one-off one. Had it indeed been a unique phenomenon, something that occurred once and never again, it is likely that its existence would never have become a part of scientific knowledge. As for phenomena that are observable by only one person and nobody else, one huge and very important class of such phenomena are phenomena related to conscious experiences, subjective experience or qualia as they are referred to by some researchers (e.g. Jackson, 1982). While we are all aware of our own conscious, subjective experiences, as we are experiencing them all the time, they are a category of phenomena that can, at this point, not be experienced by any other person but ourselves. Of course, we assume that everyone has similar subjective experiences, but no one can observe the conscious experiences of anyone but him/herself. In other words, although I know that I have conscious experiences as I can observe them all the time, how do I know that other people have them? I cannot observe conscious experiences of anyone else. And for that matter, apart from me, who and what exactly has conscious experiences? Other humans?

Animals? Plants? Inanimate objects? Conscious experiences as a category of phenomena come in conflict with this assumption of science and that is probably the main reason why their scientific status has been disputed and a subject of heated discussion for a very long time. Moreover, although conscious experiences are something that likely each of us experiences all the time, it is a class of phenomena completely unexplained by science, a class of phenomena that, according to the accepted biological and physical models should not be able to exist at all, as there is no scientific theory at the moment even remotely explaining how exactly physical processes create or transfer into conscious experiences. As David Chalmers in his 1996 book on the topic notes: "We do not just lack a detailed theory; we are in the dark about what a theory of consciousness would even look like" (Chalmers, 1996, p. ix).

This problem of conscious experiences has, however, been a topic of philosophy since philosophy began and has been studied under various names such as the mind-body relationship, the hard-soft problem of consciousness, etc. Stances researchers took related to this problem varied greatly through history and among current theorists we can see both those who consider the problem of conscious experiences to be an important scientific topic and those who would deny that the problem exists and sometimes even deny that conscious experiences themselves exist. This is also reflected in the status of conscious experiences in various social sciences. While conscious experiences are taken for granted in areas such as law and legal studies, that not only recognize conscious experiences as such, but also recognize them as causes of human behavior and utilize a number of practical methods for making inferences about them, the history of areas such as psychology has largely consisted of serious debates about the status and the very existence of conscious phenomena (e.g. Chomsky, 1959; Watson, 1913). This conflict between the nature of conscious experiences and the scientific requirement for intersubjective agreement has been the main reason why certain authors have at various points in history disputed that social sciences and particularly psychology are sciences at all. However, it should be noted that while scientists currently do not have a valid theory of consciousness, this does not mean that one will not be developed in the future nor that it will remain impossible for others to perceive individual conscious experiences. At the moment this book is written, such capabilities remain the topic of science fiction, although numerous researchers from different scientific areas recognize the issue and work on offering possible scientific explanations for the phenomenon (e.g. Atmanspacher, 2004; Gao, 2008; Hunt & Schooler, 2019).

Our mind influences the outside world only through the scientifically recognized parts of our body. This principle states that the only way we can influence the objective reality is through the interactions of our body parts with it. If we want to communicate with someone, we need to use appropriate body parts to produce a communication message the other person can perceive and understand. For example, we need to use our vocal cords to produce sounds or our fingers to type a message on a computer. If we want to move an object, we need to use our hands or another body part to lift it or move it or operate the equipment that will move it. This also means that types of interactions with the outside world that are not conducted via recognized parts of our body or organs are considered scientifically impossible. Telepathy and telekinesis, for example, are thus impossible because we have no organs or body parts that can produce the effects these phenomena would consist off. This principle also implies that our thoughts by themselves have no effect on the world unless we act upon them. Wishing someone good or bad, without doing anything to enact these results other than thinking will not produce any effects. That task we keep thinking about will not get done unless someone goes and does it. The belief that our thoughts alone can have an effect on material world is a form of magical thinking and this principle states the exact opposite of such beliefs.

Spacetime homogeneity. One of the key tenets of the scientific method is the requirement for replicability. Scientific findings need to be replicable i.e. if one person reports that certain actions under certain conditions produce certain results, other people who perform the same actions under the same conditions should obtain the same results. Textbooks on scientific methodology regularly specify that time and place by themselves are not factors that can influence events and cannot be causes of anything. Causal factors can be different processes that develop in time, but not time itself. The same stands for space. All of these expectations rely on the **belief that spacetime is homogenous i.e. that all points in space and time are the same and that the universe operates in the same way everywhere and at all times**. While this is something that most scientists very firmly believe in, it should be noted that it is not something that has or can (at present) be really and extensively tested – all scientific data currently in existence has been collected within a very narrow timeframe (when compared to the likely total age of the universe and the time in which it will likely continue to exist) and in a really tiny volume of space i.e. on Earth and its vicinity and in space it had occupied. And even this is only on certain points on Earth where the actual collection of specific data has taken place (for example at places scientific observations are carried out). In spite of this, scientists typically maintain a very firm belief that spacetime is homogenous and that all scientific laws inferred from scientific observations done will be equally valid everywhere in space and for all eternity. If we look at any scientific theory or generalization, we will see that not a single one specifies any time or space coordinates in which they are valid. Indeed, without the assumption that spacetime is homogenous, science as we know it would likely not exist and would need to be completely different. Without spacetime homogeneity, any regularities observed by scientists would have to be tested against different points in spacetime. We would not know whether regularities we observe now have been in place in previous times or whether they will work in different places. If the same actions produced different results in two different trials, we would not know whether this is because laws of nature in the two points of spacetime where actions were undertaken are different or because we made some mistake or did not account for all the factors. Making any inferences about past events would be much more difficult and inferences about distant past or future, or distant places would be completely impossible without first creating a profound understanding of how different points of spacetime differ and why. Any scientific theories or generalizations about laws of nature would have to also include specific constraints about the spacetime coordinates where they are considered to be valid. Predicting the future would also be much harder or impossible because we would not know whether the current laws of nature will remain in place in the future and for how long exactly. Luckily, the assumption of spacetime homogeneity spares science from all this and allows us to generalize any regularities we observe to all times and all places. However, a scientist should always bear in mind that the assumption of spacetime homogeneity is not something science has really been able to test so far to any great extent, nor can it really be tested to any great extent, at least with the current capabilities of the Earth civilization.

Universe is governed by a set of simple rules. One way in which we can describe the goals of science is to state that science is an effort to understand the rules by which nature and the universe function. Indeed, what almost every fundamental researcher in any field of science does is try to identify rules that govern the phenomena that are the topic of study in his/her field. But how complex are the rules we are looking for? Are they something very, very simple or are they very complicated or complex? A well-known example of this conundrum we have in the case of physics where Newtons rather simple laws of motion, were replaced by Einstein's much more complex relativity laws, which better describe and predict how universe really behaves. In the area of social sciences, we can also observe changes in the complexity of explanations. For example, psychology more or less started with sets of models called mental chemistry that tried applying explanatory models similar to those in chemistry of that time (e.g. Gentner & Grudin, 1985), over psychodynamic models presenting human psychological life as something similar to a steam boiler of the early 20th century, to the modern day psychology that tries to describe psychological functioning through linear structural equations. However, the chosen types of explanatory models, at least in the area of social sciences, has so far rarely been the result of deep understanding of the phenomena to be described and hence rarely chosen because it was seen as the most adequate for their explanation. What was more often, and typically, the case was that social scientists chose explanatory models based on what was available – the 19th century saw rapid development of the area of chemistry, hence social scientists tried to copy that. It can be argued that modern passion of social scientists for models based on linear structural equations (applied through regression analysis, factor analysis, SEM etc.) is motivated by nothing else than the easy availability of commercial statistics software that incorporate statistical procedures based on linear structural equations. Indeed, a review of the literature in the area of social sciences can hardly yield any results that include a discussion of why such equations should be chosen as basis for explanatory models or why they are considered to be the best.

Discussing this issue, an obvious question comes to mind – if these explanatory models in social sciences are chosen for their convenience rather than their adequacy, how would the best possible explanatory model look like? This question also remains valid for all other areas of science and not just for social sciences and it has given rise to what is now an area of science called complexity theory (Byrne, 1998; Manson, 2001). An important issue when discussing complexity is the question of what complexity is and how to differentiate between various levels of complexity of a system. To this end, a very useful contribution is the concept of algorithmic complexity, also known as Kolmogorov complexity or Kolmogorov-Chaitin complexity (Chaitin, 2005, 1974; Delahaye & Zenil, 2008). Algorithmic complexity is defined "as the simplest computational algorithm that can reproduce system behavior"(Manson, 2001, p. 404). In simpler words, a system whose behavior can be reproduced or recreated by using a short algorithm is considered to be simple. On the other hand, a system the behavior of which cannot be described in anything simpler than listing and describing all specific events that constitute its behavior is considered to be complex. For example, if our system was a string of characters going like this.

ABABABABABABABABABABABABABABABABABABABABAB.... and so on to infinity, we could describe this system as stating that it is a string AB reproduced an infinite number of times. Although the string we are describing is infinite, we are able to recreate it using a very simple instruction. Therefore, such a string is simple. On the other hand, imagine a string of characters where elements do not follow any rule at all, and due to this there is no simpler way to reproduce the string than writing it down such as it is. Such a string, one that cannot be reproduced in any way other than writing it, such as it is, would be considered complex. If we compare these two cases, we can note that the instructions for writing the first string are much shorter than the string itself. On the other hand, instructions for writing the second string are at least as long as the string itself, making the string complex, actually, as complex as a system can be.

Now if we try to apply the same reasoning to the functioning of the universe, we can note that the goal of science is to develop a set of instructions that can describe the functioning of the universe, while at the same time being a part of that universe. The question of whether this is possible arises. This is the question of whether the rules governing the universe we live in are simple enough that instructions needed to reproduce them can be contained in the universe itself or are sufficiently complex that instructions needed to reproduce the behavior of the universe would be too large to fit inside the universe. In the first case, the goal of science of creating a perfect model of the universe would be achievable, in the second case, it would not be possible. But which one is it? Is our universe simple or complex? On the one hand, we can observe that in many areas of science we have so far been able to reasonably well predict the behavior of huge natural systems through the use of relatively simple equations and mathematical models. This can be taken to indicate that the true complexity of the universe is not near the maximum and that it is likely somewhere in the lower degrees of complexity. But how simple? Some authors, like, for example, Wolfram have proposed the idea that the true complexity of the universe might be very low i.e. that our universe is actually so simple that it could likely be reconstructed through a small set of very simple rules (Wolfram, 1984). However, this approach has so far not seen much development beyond the initial general idea and due to this has received criticism (e.g. Chaitin, 2005; Manson, 2001). Although the considerations about complexity are still relatively new in science in general and social scientists rarely discuss and debate the true complexity of the phenomena they study, we can conclude that scientists implicitly believe that phenomena studied by science are sufficiently simple to enable the existence of science i.e. simple enough that they can be reproduced using a model (or sets of instructions) that are sufficiently small to fit into our universe. Otherwise, science as such would likely not be possible. Given this situation, it would also be recommendable that scientists, especially in area of social sciences routinely consider the likely complexity of rules governing phenomena they study, instead of blindly building science based on whatever types of mathematical models are available in the statistics software they use. That said, it should be noted that the concept of complexity is already making some inroads into the field of social sciences (e.g. Gauvrit et al., 2017).

Time is a special dimension – phenomena only move forward through time. Although we accept that spacetime is homogenous, that all points in space and time are the same and governed by the same laws of nature, time is special in a way that phenomena can only move through time in the forward direction. Unlike dimensions of space in which phenomena can move in all directions, phenomena can move only forward through time (as far as scientific research outside the area of physics is concerned). Due to this, science recognizes only the existence of cause-and-effect relationships in which cause precedes the effect in time or in which cause and effect happen simultaneously. Causes cannot produce effects in the past, only in the future. Most epistemological discussions about cause-and-effect relations explicitly list the requirement that cause precedes the effect in time or that they are simultaneous as one of the key requirements for determining that a certain relationship indeed is a cause-and-effect relationship. In other words, we assume that past can influence the present and the future, but that future cannot influence the past. For example, one of the most prominent philosophers of the modern age – David Hume in his famous work "A treatise of human nature" states that in order for some X to be a cause of Y, X must temporally precede Y. This view is also shared by another famous philosopher of science – John Stuart Mill and most of the later scientists who studied the phenomenon of causation. It is also one of the key principles for determining causation that can be found in methodology textbooks and literature dealing with experimental procedures (e.g. Milas, 2009).

Knowable causes. The phenomena we are observing have causes that lead to their appearance and it is in principle possible to determine what those causes are. Not only are the rules that govern the universe simple enough for the universe to be able to contain a sufficiently precise model of itself, an assumption of science is that we are capable of knowing what the causes are, capable of understanding the set of rules that governs the universe. This is a principal belief that gives meaning to scientific research. If this was not the case, if it was not possible for humans to discover and understand the network of causes and effects of the universe, scientific research directed at this would be in vain. Also, if we are at present not able to identify the cause (or causes) of a certain phenomenon, this belief requires us to conclude that this is because we have not studied it adequately and not because the causes of the phenomenon are unknowable in principle. We just have to continue studying and looking for causes and when we find the right approach, methodology or measurement instruments, we will be able to discover those causes, because all of the causes are knowable. There are no phenomena causes of which cannot be discovered. In philosophy, a variant of this principle is known as the principle of sufficient reason and is defined as the belief that "everything must have a reason, cause, or ground" or in other words that "For every fact F, there must be a sufficient reason why F is the case."(e.g. Melamed & Lin, 2021).

1.2 The nature of the causal network of the universe – stochasticism vs. determinism

One other important aspect of how we build the scientific understanding of the world around us is the **view about the nature of the causal network of the universe i.e. about the relations between causes and effects**. And it is this view of the nature of the cause-and-effect relationships in the universe in which various scientific areas de facto differ and where we also often see two not quite compatible views of the universe used interchangeably in scientific work. One of these worldviews is called determinism (e.g. Hoefer, 2016; James, 1884) and the other one, opposite to it, we will call stochasticism.

The principle idea of **determinism**, causal determinism or "hard" determinism as some call it, is that the universe is a very complex network of causes and effects, such that every event is completely determined by a set of causes. Each of these causes is in turn completely determined by another set of causes and the network extends to eternity or to the beginning of the universe, if such a point existed. This means that if we knew all the starting conditions of the universe and all the laws of nature, we would be able to predict without any errors all events that would take place in the universe through its entire existence. An assumption that our universe is a deterministic one i.e. a deterministic view of the universe has some very important implications:

- Since the universe is a deterministic mechanism of causes and effects with causes completely determining effects, any inability on our part to predict certain events with total accuracy is due to our insufficient knowledge of natural laws and causes of the event and not due to the event being unpredictable in principle. If we improve our knowledge, we will be able to make totally accurate predictions of everything.
- For the same reason, everything that will ever happen in the universe has been predetermined by the conditions at the beginning and there is no way to influence the course of events. Fate exists and we cannot change it. What will happen, will happen, no matter what we do.
- This worldview also implies that there is no free will or that free will is merely an illusion because everything is predetermined.
- Taken together with other assumptions of science, most of all the assumption that the rules governing the universe are sufficiently simple, that a perfect model of the universe can fit inside the universe, this worldview also implies that it is possible to develop a so-called theory of everything, a perfect model of the universe that will totally accurately describe and explain the functioning of the universe.

The question of whether the nature of our universe is deterministic has been the topic of philosophical debates for centuries with philosophers discussing various variants of the deterministic worldview that included such things as "soft determinism" that would allow for free will in contrast to the one described here, that would be termed "hard determinism" (e.g. James, 1884). Scientists outside the area of philosophy rarely openly talk or declare the worldview they have when creating scientific theories and explanations, but from how theories are typically formulated, especially in the area of natural and technical sciences, it can be concluded that a majority of scientists in these areas considers determinism to be the true nature of our universe regardless of whether they are aware of this philosophical dilemma or not. It should also be noticed that in a deterministic universe, there is no place for random events, hence there is no place for assumptions the science of statistics is based on, as we will see later.

The principle idea behind **stochasticism** is that our universe is stochastic in nature meaning that stochastic processes can and do exist, processes where the same set of causes under the same set of conditions can lead to a multitude of different consequences, each having its own probability of occurring. A stochastic universe is a universe in which there exist outcomes that cannot be accurately predicted even when we know all the causes and all the laws of nature, events that are unpredictable in principle. This is a universe in which events are not predetermined and knowing the starting conditions of the universe does not allow us to accurately determine what the ending conditions will be. The notion that our universe might be stochastic in nature has also some very important implications:

- Not being able to predict something with sufficient accuracy or at all does not necessarily mean that we do not know enough. It is possible that the nature of the phenomenon in question is such that it cannot, in principle, be predicted at all, since the same process can lead to different outcomes in a stochastic universe, because that is the nature of the universe.
- For the same reason, the way how events in the universe will unfold cannot be predicted with total accuracy or cannot be predicted at all. Since a same set of causes can lead to different outcomes, we cannot know for sure which outcomes will

happen in each particular case. Knowing the starting conditions of the universe and all the laws of nature is not sufficient to be able to predict the end conditions of the universe nor what exactly will happen in the meantime. The universe is essentially unpredictable. This also means that there is no fate and events are not predetermined.

- It is possible that something like free will might exist in some forms of a stochastic universe, depending on the nature of these stochastic processes. However, stochastic universes in which there is unpredictability without free will can also be imagined. This largely depends on what the nature of free will might be, as it is a phenomenon science currently has no sufficiently good model of.
- Since the universe is unpredictable, no matter how much we learn about it, we will never achieve a model of the universe that allows us to totally accurately predict events in the universe. The best we can do is assess probabilities of certain outcomes and even this might be a vain endeavor as these probabilities might be ever changing, depending on the specifics of the stochastic nature of the universe.

While not many scientific theories and generally theorists will explicitly state that they believe that our universe is a stochastic one in the meaning described above, there are many approaches in various areas of science that practically rely on methodology and scientific tools designed for a stochastic universe. The most widely used of such scientific tools is the concept this book is all about – statistics. We can find statistics being the basis of scientific research in all areas of science ranging from natural sciences such as physics, especially quantum physics, biology, particularly in areas such as genetics, in medicine (population studies, epidemiology etc.), chemistry etc. to social sciences. If one looks into publications presenting results of empirical research in modern economics, psychology, sociology, education, actually any other social science, one will see an abundance of statistics and an extreme reliance on stochastical models. This is the very reason why this distinction between a deterministic and a stochastic worldview is important for a book on statistics as it represents **a very large and important contradiction in modern science** that all working in it need to be aware of. In majority of cases, **we will see scientists trying to formulate a deterministic theory that explains the phenomena they are studying while using statistics i.e. scientific procedures based on the stochastic worldview, to process the data on those same phenomena**. To add to that, many will try to deny this contradiction by stating that statistical procedures are used because they are useful, which clearly is the case, and that phenomena behave in a stochastic-like way because we do not know sufficiently about them. They might believe that treating phenomena like they are stochastic in nature is the first step in studying them and that, after we have learned enough, we will be able to create accurate deterministic models, after which we will not need statistics. In this sense, statistics and procedures based on the stochastic worldview in general are seen as a necessary evil, a needed first step in studying the world, something that will be replaced by deterministic models when sufficient knowledge is obtained. This attitude can probably best be illustrated by the words of the great physicists Albert Einstein who, evaluating the then novel quantum theory in physics that adopted the stochastic approach, stated "Quantum theory yields much, but it hardly brings us close to the Old One's secrets. I, in any case, am convinced He does not play dice with the universe." (e.g. *Physics and Beyond: "God Does Not Play Dice", What Did Einstein Mean?*, 2021). These words clearly state his belief, a belief shared by likely most scientists, that, while stochastic methods are useful, they do not reflect the true nature of the universe, which is deterministic. Be that as it may, the fact remains that the true nature of our universe is, to this date, unknown and all we have are beliefs about its nature. The fact that we can predict many natural phenomena with great, even total accuracy, means only that these phenomena have been predictable so far, does not mean that all phenomena everywhere are predictable and that there are no phenomena that are unpredictable per se i.e. it is not the proof that our universe is deterministic in nature, as a stochastic world also allows the existence of sets of causes that have only a single consequence, that are predictable. The fact that we have many very good deterministic models does not mean that we will be able to create such models for phenomena for which we currently do not have them. And most important of all, it does not diminish or eliminate the fact that using procedures based on the assumption that the world is stochastic in nature to build and support deterministic theories, theories that leave no place for stochastic processes is an important contradiction of modern science. A contradiction everyone applying statistical procedures in scientific research should be well aware of.

1.3 Scientific explanations

When talking about components of science relevant for an introductory text in statistics we unavoidably come to the topic of scientific explanations. One of the key goals of science is to be able to explain the phenomena we experience i.e. to be able to give an answer to questions starting with "why". Scientific explanations and their nature have been a topic of intense discussion and study by philosophers of science for centuries and these have resulted in a number of philosophical models of scientific explanations (e.g. Woodward & Ross, 2021). An explanation is usually described as a set of statements that describe a set of facts and clarify its causes and consequences. Hempel & Oppenheim (1948) proposed that a scientific explanation consists of two parts – the explanandum and the explanans. The explanandum is the sentence describing the phenomenon that is to be explained and the explanans are the sentences that are "adduced to account for the phenomenon" (Hempel & Oppenheim, 1948, p. 137). For an explanation to be adequate, the explanandum must be logically deducible from the information contained in the explanans, the explanans must contain general laws and the content of the explanans must be empirical i.e. it needs to be testable through empirical research. While most scientists would agree that the best explanations in science are those that explain phenomena through causes and consequences and that we might even consider the end-goal of science to be the creation of a causal model of the universe, thus explaining the universe and everything in it in this way, it is a fact that many useful and widely accepted scientific explanations fall short of this type of explanation as there are many phenomena that scientists cannot explain through causes and consequences at the moment. Some of this is surely due to insufficient knowledge, but whether some of it is because we are dealing with essentially unexplainable phenomena is something we cannot tell at the moment (please refer to the discussion about the deterministic vs. stochastic nature of the world in the previous chapter, or the philosophical debates about determinism vs. indeterminism).

This means that science operates with different types of scientific explanations and for proper understanding of the topic of this book – statistics it is of key importance to understand where inferences made through statistical procedures come into play and to what types of scientific explanations they can contribute and how. One way to organize types of scientific explanations would be into following categories:

- **Causal explanations** – are a type of explanation where the phenomenon in question is explained through the specification of causes leading to it and/or the consequences it causes. Explanations of this type use references to established scientific theories, natural laws or general rules to explain the interactions of various variables and elements included in the explanation. Ideally, a phenomenon is explained by subsuming it under laws or under a theory (e.g. Hempel & Oppenheim, 1948; Hitchcock, 1975) It might be said that the end goal of science is to be able to explain everything in the universe through a network of causal explanations. This type of explanation is generally the best type of scientific explanation. However, providing a causal explanation also requires a deep, detailed and organized knowledge of the matter, knowledge to which the explanation itself will contribute and become an integral part of and this is something that is currently simply not available for many phenomena and even classes of phenomena. Causal explanations for example, include the explanation of why planets in our Solar system orbit the Sun through referring to appropriate laws of physics or an explanation of why it rains that accounts for how interaction of atmospheric conditions turns water vapor into water drops and what causes them to fall to the ground. In life sciences, that would, for example, be an explanation of how exposure of eyes to light causes impulses in the optical nerve that would involve the explanation of physical, chemical and biological properties of the eye that allow light to cause this etc.

- **Statistical or probabilistic explanations** – are explanations where something is explained or, more often, predicted with a certain probability by noting the fact that in the past the phenomenon used to appear in conjunction with another phenomenon. We have observed in the past that two types of events tend to happen together and from that we infer that this might continue in the future. We might not know why they happen together; we might not know whether one is the cause of the other; we might even know that they do not always happen together, but we have registered that when one happens it is more likely that the other also happens than not and we infer, or better said – hope, that this will continue in the future. Relevant for the concept of statistical explanations is the Reichenbach's common cause principle (C. Hitchcock & Rédei, 2020; Reichenbach, 1956) that states that when two events tend to constantly happen together (when they are probabilistically correlated) that is either because one of them causes the other or because both are caused by a third factor (or a group of third factors) that preceded in time both of these events. This means that when we make a generalization or an explanation based only on the observation that two events tended to happen together in the past, we do not know whether this might be due to both of them being caused by some other factor not observed at all. This also means that if such a factor changes in the future without our knowledge, the first two factors might stop happening together. This is something well known to people working with predicting, for example, market movements. Due to these properties, statistical explanations are usually considered to be **the worst type of scientific explanations**. However, they are also the type of explanation that **requires the least bit of theoretical knowledge and understanding of the matter.** A statistical inference about probabilities of two events happening together can be made without understanding the deeper laws of nature governing these events or without having a good theory or any theory about the phenomena in question. This makes statistical explanations usually the first types of explanations and inferences to be made about previously unknown phenomena, as they can be made purely based on empirical observations. It can be said that the development of scientific understanding starts with statistical explanations and progresses towards causal explanations as researchers learn more about the studied phenomenon and develop systematic knowledge about it. This book covers statistical concepts and procedures necessary to make statistical inferences that allow the creation of this type of explanation. That is why it is necessary for the reader to understand the position of statistical observations in the typology of scientific explanations and the fact that they are the worst but also the first type of scientific explanation that can be created, but also that creating anything better than pure statistical ones requires things other than statistics i.e. for other types of explanations statistical data and statistical procedures are not sufficient.

Apart from these two types of explanations, the, conditionally, best type – causal and the worst type of explanation – statistical, science also applies other types of explanations, the epistemological value of which can be considered to be somewhere in between, explanations that on one hand rely on more knowledge than pure statistical/probabilistic explanations, but that, on the other hand, still fall short of causal explanations, unable to explain causal relations between phenomena being explained. These types include:

- **Developmental explanation** (e.g. Kitchener, 1983; Woodward, 1980) – also referred to as genetic explanations (Nagel, 1961), are a type of explanation where we explain why a system is in a certain stage of development by referring to a law that describes a sequence of stages through which systems of that type go through (Kitchener, 1983). Woodward (1980) proposes that there are actually three types of developmental explanations – a) when the behavior of a certain organism is explained by reference to the developmental stage of the organism (for example explaining the crying of a baby by listing the fact that it is a baby, a human individual at ane early stage of development); b) when explaining that an object or an organism is in a certain stage or will enter a certain stage by referring to the law about the sequence of stages in the development of that object or organism and c) when we explain the law about the sequence of stages by referring to a more general developmental principle or mechanism of progression through developmental stages. Developmental explanations are found in all areas of science, but probably the most well-known are those found in biological sciences explaining the stages of development of organisms or those in psychology describing psychological development of individuals through childhood and lifetime.
- **Functional explanations** – are explanations in which a unit is explained by specifying the function it serves as part of a larger system, usually by specifying the trait or property of the system that the existence of that particular unit allows. An example of this type of explanation would be when we explain the role of the heart in the human body by specifying that the function of the heart is to pump blood through our system of blood vessels thus enabling the circulation of blood through the body. An example from social sciences would be explaining the existence of schools in a society by stating that they exist in order to educate children, thus contributing to the general education level of the society.

• **Teleological explanations** of the behavior of living organisms – are a specific type of explanation where the behavior of a person, group of persons or of a living organism in general is explained by specifying the goals of such a behavior. For example, we can explain the fact that a student is studying by specifying that he/she has the goal of passing a course (for which he/she is studying). In this way, we have explained the behavior by stating a goal or purpose a person has and due to which that person behaves the way he/she behaves. While some authors consider functional and teleological explanations to belong to the same category of explanations (e.g. Nagel, 1961), it is our belief that there is a substantial difference between an explanation where a researcher explains the role that a certain part of a wider system plays in the functioning of the system (such as in the situation of the heart pumping blood through the body, or of atmosphere providing a breathing medium for living organisms thus allowing the Earth's ecosystem) and a situation where we explain intentional behavior of humans and living organisms through specifying their intents, such as is the common situation, for example, in psychology and in legal studies when intent is an important factor in describing and evaluating the behavior of a person.

1.4 Self-fulfilling prophecies and the corruption of statistical indicators

When discussing laws, generalizations and predictions of science in the context of behavior of humans, it should be noted that applying knowledge of these in making decisions to be applied through practical interventions can lead to changes in behavior of people subject to those decisions and interventions that are either in line with the expectations of those making the rules or intent to subvert or oppose them. This happens because when people are aware of the concepts those in power over them use to make decisions and those decisions are important enough for them, they will form an active attitude towards those decisions and interventions and might decide to change their behavior accordingly. This can sometimes result in so-called **self-fulfilling prophecies.** Self-fulfilling prophecies happen when people, who might otherwise behave differently, start behaving in a way that is expected of them after they find out and start believing that that is the behavior expected of them. For example, a student who was not good at his studies but was trying to improve, might give up trying and "accept his/her fate" after being told, after, for example, cognitive testing, that his cognitive capacities are poor, thus fullfiling the expectation that students with poor cognitive capacities will fare poorly. A person thinking about starting a new business, might give up on the idea after being told that the economy of his area will stagnate. In this way future contribution of his business to the economy will be eliminated, thus indeed fulfilling the expectation that the economy will stagnate.

Self-fullfiling prophecies are also an issue with making prediction of stock market movements – statements of expectations of various analysts and important figures about how price of stocks of companies will move in the future, might intice people to buy or sell these stocks, thus making the expectation that the price will move in that direction happen. It is a known strategy of "shorters" (people short-selling i.e. selling borrowed stocks with the expectation to buy them back when their price drops) to try to form public expectation that the price of the stock they shorted will drop by creating negative publicity for the company whose stock they are shorting. People who were following the US stock market movements in the late 2020/early 2021 could have witnessed how the price of shares of one company on a visible and protracted downtrend in business increased more than tenfold in a matter of months and without any business performance related news to justify that, just due to expectations of people formed through social media and advertising. This particular event is now becoming known as the "Gamestop frenzy".

Another example of self-fulfilling prophecies are various **placebo effects**, situations where ineffective treatments manifest certain effects because people these treatments are applied on are made to expect these effects. A very well-known example of this type is the so-called **Hawthorne effect**, named after a series of studies on the productivity of workers under various physical conditions conducted in the early 20th century. In these studies, researchers examined how different working conditions, such as e.g. illumination, work pauses or working hours affect worker productivity. Results showed that productivity of groups of workers involved in the study generally improved compared to regular workers not included in the study more or less regardless of what the researchers did. This was attributed to the workers knowing that the researchers were researching productivity. For example, when researching effects of illumination, the productivity of workers was increasing even when researchers started decreasing illumination ever more and only after it was decreased to the level of moonlight at night did the workers start to complain and a similar thing happened with other variables (e.g. Wickstrom & Bendix, 2000).

Another issue concerning the application of generalizations in social sciences for the purposes of decision-making and interventions is the phenomenon of **the corruption of statistical indicators.** In public administration, company management and various other organizational settings, people in charge of managing these organizations often rely on various ways of measuring developments of importance and base their decisions on them. For example, for assessing cognitive abilities, certain tasks are administered to people in the scope of psychological tests; for measuring inflation, prices of certain commodities are followed in order to calculate a price index; in order to assess literacy or education quality of a country, pupils from certain schools are tested; in order to monitor the changing of stock market prices, prices of stocks of certain companies are followed. Teachers might examine the progress through the curriculum by examining the knowledge of only select parts of the curriculum and not of all of its parts. Then, decisions are being made based on the values of these indicators, decisions affecting various people and in which various stakeholders might be interested. **Corruption of statistical indicators happens when those affected by certain indicators and/or those with interests in outcomes of decisions based on these indicators, understanding how the indicators work and how they are calculated, take action to change the value of indicators to their benefit, but just of the indicators, not of the underlying property the indicator is supposed to assess.** When this happens, the affected indicators will become invalid and this might lead to the decision-makers making invalid decisions based on the now-nonvalid indicators. For example, students knowing what exactly their teacher will ask them in the exam might learn just those things while completely neglecting the rest of the curriculum and thus receive an excellent grade in spite of knowing very little. i.e. nothing apart from the specific questions they learned. In a similar fashion, a government interested in showing that inflation is low, might take action to reduce the official prices of specific commodities that are included in the index (even at the cost of causing a shortage), while letting the prices of other commodities explode, thus producing a low inflation index in spite of a rampant inflation. A company publicly traded on a stock exchange might decide (as various famous real world examples readily show) to order its own products from itself and than pay it back to itself in order to inflate its revenue statistics. The investors would then conclude that the company is succesfully growing its business (as indicated by increasing revenue) and would be more willing to buy the shares of the company, thus increasing their price on the stock market. One very famous historical example of the phenomenon of the corruption of statistical indicators is the affair of steel production in the communist China during the Great Leap Forward (https://en.wikipedia.org/wiki/Great_Leap_Forward) in the mid-20th century, when inhabitants of various villages were ordered to produce steel, without having the resources or the infrastructure and equipment necessary for industrial quality steel production. This resulted in these villages producing completely useless "pig steel" from various available objects containing steel for no other purpose than recording that production of steel in China has increased. At the time, steel production in a country was used and as an indicator of the development of a country, as industrial, and thus developed, nations of the time produced much more steel than nondeveloped ones because steel was needed for infrastructure development and the production of various commodities. Of course, it soon became obvious that spoofing a statistical indicator of development is not the same as the actual development and the consequences were dire.

Probably the widest and longest lasting example of this effect was the so-called Flynn effect in intelligence and cognitive testing. Namely, it was noted that, during the 20th century and somewhat into the earliest years of the 21st century, achievements of people throughout the developed world on intelligence tests was increasing (e.g. Hedrih, 2020). While psychologists posed various hypotheses to explain the phenomenon (e.g. Sundet et al., 2004; Teasdale & Owen, 2005), in the end it turned out, as explained by Flynn himself (Flynn, 2007) that, as results of intelligence tests and achievements on tasks similar to those in intelligence tests, were more and more used for making various decisions that were important for people's lives, people became more and more familiar with the tasks like those found in intelligence tests and increased their proficiency in solving them, while intelligence, the trait these tests were supposed to measure, remained more or less the same on the population level. In this way, the acquisition of a relatively useless skill (proficiency in solving tasks akin to those in intelligence tests) corrupted/invalidated indicators of important cognitive abilities, forcing psychologists to innovate and reformulate the rules for interpreting test results in order to maintain their validity.

All of these examples show why both the issue of self-fulfilling prophecies and of the corruption of statistical indicators need to be taken into account when using scientific explanations, theories and similar vehicles in the area of social sciences for making decisions that affect people and why it is necessary to be constantly aware that applying certain rules on people might lead to people changing their behavior in response to the new rules and decisions. We should also be aware that many of the regularities found in behavioral sciences hold only because people keep perceiving the relevant situations in a certain way and decide to behave accordingly and that this is something that may change if the situation changes as a result of intervention based on those very regularities. Therefore, unlike the situation in the natural sciences, where inanimate objects do not change their behavior based on our decisions, applying the findings and generalizations of social sciences on people might often make those very same findings and generalizations no longer valid.

1.5 Science and pseudoscience

It can easily be said that modern world has been built on the foundations of science. Everywhere around us, most of the objects we are using, objects we depend on for maintaining our way of life or our life itself are products of science, most of them the end results of centuries of scientific research and development. Due to this, science has great authority in our society and people trust it daily with their very lives. This trust in science is earned because it is the result of concerted efforts of scientists all over the world rigorously using the scientific method and testing, examining and reexamining their results and findings. However, as probably with all things valuable, there are those who would like to use that trust for their own goals and pushing their own agenda which is typically incompatible with science, but is aimed at serving their personal interests, usually at the expense of other people. And this is how pseudoscience comes to be. We can define **pseudoscience as any set of beliefs, claims or procedures that are attempting to be presented as real science, but are not obtained through the use of the scientific method, that lack evidence or logical consistency, that cannot be verified or falsified (Popper, 1963) or that for other reasons cannot be considered scientifically valid**. While this definition of pseudoscience seems clear on the abstract level, the question of practical demarcation between science and pseudoscience is an important one. If done incorrectly it could lead either to classifying genuine and valuable scientific efforts as pseudoscientific or to the inclusion of pseudoscientific practices and systems into science thus devaluing and even derailing real science (e.g. Lakatos, 1978).

So how do we differentiate between science and pseudoscience. The famous 18th century philosopher David Hume is attributed saying: "If we take in our hand any volume; of divinity, or school of metaphysics, for instance; let us ask, does it contain any abstract reasoning concerning quantity or number? No. Does it contain any experimental reasoning concerning matter of fact and existence? No. Commit it then to the flames. For it can contain nothing but sophistry and illusion" (Lakatos, 1978, p. 1). While this is essentially true, its practical application is quite a different thing. Namely, while there are clearly pseudoscientific systems that involve numbers and abstractions (e.g. numerology, astrology), there were also situations through the history of science where legitimate scientific endeavors were considered pseudoscientific by some, such as was the case with psychology (e.g. Ferguson, 2015; Skinner, 1990) or with the scientists adopting Mendelian genetics who were at a point treated as pseudoscientists in the USSR (Lakatos, 1978). On the other hand, there were also times when clearly pseudoscientific systems were accepted as science, such as was the case of the 17th century witchcraft being treated as a model scientific approach by the house philosopher of the UK Royal Society (Lakatos, 1978) or for example the modern acceptance by various governments of "cosmic energy medicine" methods as scientific. All this said and being aware that the so-called question of demarcation between science and pseudoscience was and still is a subject of great debate among philosophers of science and scientists in general, it can be noted that there are some tendencies and regularities that can be used in practice to help tell whether what we are dealing with is real science or pseudoscience posing as science. These differences can be seen in:

- **Publishing results** – scientists primarily first publish their results in scientific journals and similar scientific publications that are reviewed by other experts in the field, aimed at expert audiences and that maintain strict standards of scientific honesty and accuracy. Only after the results have successfully passed evaluation and intense scrutiny by other experts, including primarily those independent of authors of the results, will they be presented to general public, usually in forms of secondary publications created for the general public. Unlike this, pseudoscientists aim their publications primarily at the general public, preferably people without sufficient competencies and knowledge to adequately evaluate the materials of pseudoscientists, while markedly avoiding experts, often explaining this behavior through "avoiding conspiracy" against them "organized" by experts or stating that their publications are banned (when they are not) and similar. Pseudoscientific publications do not undergo reviews by experts in the field, do not undergo verification and there are no demands in regard to accuracy or veracity of claims included in them. Authors and distributors of such publications will try distributing them in your local barber shop or in a caffe or selling them to naive people, but they will actively avoid scrutiny or evaluation by experts.

- **Replicable results** – science requires replicable results. Due to this, scientific procedures will be (in scientific publications and reports) explained and described in sufficient detail to enable others to exactly repeat the same procedure and to verify for themselves whether equal results will be obtained. On the other hand, pseudoscientific "results" can neither be replicated nor verified. If there is any description of the "research" procedure at all, it is only vaguely described, so that it is impossible to definitely determine what exactly was done and how. For example, one relatively widespread pseudoscientific system describes the origin of its knowledge through a story about how the system was founded by a couple of people who discovered all elements of the system because "they were very talented people with ingenious insights" and explicitly deny that their research procedures could be repeated by anyone else, because no one else has the talents and "special abilities" of their founders. Whatever you do to replicate a pseudoscientific "research procedure", if your results are negative, the pseudoscientist will tell you that you made some mistake and did not replicate the procedure correctly! Of course, such approach has very little in common with the scientific approach.
- **Treatment of errors** – in science, errors are actively sought. When one group of scientists conducts a study and publishes results, other scientists will review their findings and, in this process, actively search for any errors that were made, alternative explanations that were missed or any other irregularities. Scientists know that even theories that are false can sometimes accidentally yield correct predictions. That is why they will test the assumptions of scientific theories in new situations and under new conditions, knowing that if the theory is correct it will not make false predictions. On the other hand, pseudoscientists ignore errors, justify them, lie about them, reject them, explain them away, forget about them, try to hide them and do whatever they can to minimize and divert attention away from them. This is because the goal of pseudoscientists is to convince others that their claims are true, not to really test the veracity of their claims.
- **Progresses of knowledge** – science progresses, as more and more research studies are conducted, scientists learn more and more about the phenomenon studied and that means that the physical processes underlying the phenomenon studied become ever better studied and knowledge about them increases. Scientific research of a phenomenon may start from scratch, but every new study increases the amount of

knowledge. Contribution to scientific knowledge is a basic requirement when evaluating any scientific work. Scientific work that does not contribute new knowledge, either through verification of the existing generalizations and theories or through new discoveries, is generally considered worthless. In contrast to that, pseudoscience does not progress and nothing particularly new is learned in time. The physical process underlying the phenomenon is not explored nor researched. Whatever is known when the system is created generally remains the same throughout the existence of the pseudoscientific system.

- **Reliance on evidence** – science convinces people by presenting evidence, by using logical arguments or mathematical considerations, models and descriptions. It tries to make the best of the available data. When new evidence become available that show that previous interpretations were not correct, this is recognized, these interpretations are abandoned and new ones are sought. On the other hand, pseudoscience tries to get people to believe in its claims by

Table 1.1 Practices that tell whether what we are dealing with is real science or pseudoscience posing as science.

Criterion	Science	Pseudoscience
Publication of results	First published for experts, undergoes review and intense examination and scrutiny by other experts. Published for general public only after it successfully passes evaluation by experts	Targets general public, usually those without competences necessary to evaluate it. Review or evaluation by experts is actively avoided.
Replicable results	Science requires replicable results, procedures are described in sufficient details for other scientists to be able to replicate them and thus verify the results themselves	Results cannot be verified. Procedures are not described or are described only vaguely so that they cannot be replicated.
Treatment of errors	Errors are actively sought in order to correct them.	Errors are ignored, hidden, explained away, lied about etc. Initial concepts are never abandoned are corrected.
Progress of knowledge	Science progress. With new studies the quantity of knowledge about the phenomenon and its underlying physical processes increases. We learn more with each new study.	There is no progress with time. A pseudoscientific system remains more or less the same as it was when it was created. New "studies" do not produce new knowledge.
Use of evidence	Science convinces through presenting evidence, logical arguments and mathematical considerations, functions and descriptions.	Pseudoscience tries to get people to believe in it by relying on faith, trying to convert rather than convince. It demands that people believe in it in spite of the existing evidence, not because of it.
How it is supported	Does not sell untested or fraudelant products. Scientific products undergo r rigorous examinations and testing before being placed on the market	Sale of problematic products is the main source of income. Great promises are made in advertising products, but the products do not deliver, have either little or no effect and can sometimes be harmful for the user.

relying on "faith" and by demanding that people simply believe their claims without questioning them. They tend to try to convert people by adopting a pseudoreligious character to their beliefs and get them to believe in them in spite of facts showing that these beliefs are false. Pseudoscience never abandons its initial ideas no matter what the facts say.

- **How it is supported** – science places on the market only rigorously verified procedures and products. Every product of science undergoes a rigorous and thorough testing process before it is offered to the public. In this process, both its safety and effectiveness are established. If the product does not pass these tests and cannot be improved sufficiently to pass them, it is abandoned and never reaches the market. On the other hand, the main source of income for pseudoscience and pseudoscientists it the sale of various problematic products, products that come with great promises but, in better cases, deliver little or nothing or, in worse cases, even cause harm to the buyer. Such products might include ineffective "medicine", books and "educational" courses making promises of great results or pseudoscientific services such as fortune telling, relying spiritual messages, horoscopes etc.

References

Atmanspacher, H. (2004). Quantum Theory and Consciousness: An Overview With Selected Examples. *Discrete Dynamics in Nature and Society*, *1*, 51–73. 10.1155/S102602260440106X

Baldwin, T., & Bell, D. (1988). Phenomenology, Solipsism and Egocentric Thought. *Proceedings of the Aristotelian Society, Supplementary Volumes*, *62*, 27–43.

Byrne, D. (1998). *Complexity Theory and the Social Sciences: An Introduction*. Routledge.

Cen, J., Xue, S., & Groningen, H. (2014). Observation of the Optical and Spectral Characteristics of Ball Lightning. *Physical Review Letters*, *112*(035001), 035001-1-035001–035005. 10.1103/PhysRevLett.112.035001

Chaitin, G. (2005). Epistemology as Information Theory: From Leibniz to Ω $\star$.

Chaitin, G. (1974). Information-Theoretic Computational Complexity. *IEEE Transactions on Information Theory*, 10–15. https://pdfs.semanticscholar.org/9a4f/05562a59842ce2107e2e858c8c8a7302329c.pdf

Chalmers, D. J. (1996). *The Conscious Mind: In Search of a Theory of Conscious Experience*. Oxford University Press.

Chomsky, N. (1959). A Review of B.F. Skinner's Verbal Behavior. *Language*, *35*(1), 26–58. http://cogprints.org/1148/1/chomsky.htm

Delahaye, J.-P., & Zenil, H. (2008). Towards a stable definition of Kolmogorov-Chaitin complexity. *Fundamenta Informaticae*, *XXI*, 1–15.

Ferguson, C. J. (2015). "Everybody knows psychology is not a real science": Public perceptions of psychology and how we can improve our relationship with policymakers, the scientific community, and the general public. *American Psychologist*, *70*(6), 527–542. 10.1037/a0039405

Flynn, J. (2007). *What is Intelligence? Beyond the Flynn Effect*. Cambridge University Press.

Fodor, J. (1991). Methodological Solipsism Considered as a Research Strategy in Cognitive Psychology. In R. Boyd, P. Gasper, & J. D. Trout (Eds.), *The Philosophy of Science* (pp. 651–669). The MIT press. http://www.cog.brown.edu/courses/cg2000/Papers/Fodor1991PhilSci.pdf

Gao, S. (2008). A Quantum Theory of Consciousness. *Mind & Machines*, *18*, 39–52. 10.1007/s11023-007-9084-0

Gauvrit, N., Zenil, H., Soler-Toscano, F., Delahaye, J.-P., & Brugger, P. (2017). Human Behavioral Complexity Peaks At Age 25. *PLOS Computational Biology*, *13*(4), 1–14. 10.1371/journal.pcbi.1005408

Gentner, D., & Grudin, J. (1985). The Evolution of Mental Metaphors in Psychology: A 90-Year Retrospective. *American Psychologist*, *40*(2), 181–192.

Hedrih, V. (2020). *Adapting Psychological Tests and Measurement Instruments for Cross-Cultural Research: An Introduction (1st Edition)*. Routledge, Taylor&Francis Group.

Hempel, C. G., & Oppenheim, P. (1948). Studies in the Logic of Explanation. *Philosophy of Science*, *15*(2), 135–175.

Hitchcock, C. R. (1975). Causal Explanation and Scientific Realism. *Erkenntnis*, *37*(2), 151–178.

Hitchcock, C., & Rédei, M. (2020). Reichenbach's Common Cause Principle. In E. N. Zalta (Ed.), *The Stanford Encyclopedia of Philosophy (Spring 2020 Edition)*. https://plato.stanford.edu/archives/spr2020/entries/physics-Rpcc/

Hoefer, C. (2016). Causal Determinism. In *The Stanford Encyclopedia of Philosophy* ((Spring 20). https://plato.stanford.edu/archives/spr2016/entries/determinism-causal

Hunt, T., & Schooler, J. W. (2019). The Easy Part of the Hard Problem: A Resonance Theory of Consciousness. *Frontiers in Human Neuroscience*, *13*, 378. 10.3389/fnhum.2019.00378

Jackson, F. (1982). Epiphenomenal Qualities. *The Philosophical Quarterly*, *32*(127), 127–136.

James, W. (1884). *The Dilemma of Determinism*. Kessinger Publishing.

Kitchener, R. F. (1983). Developmental Explanations. *The Review of Methaphysics*, *36*(4), 791–817.

Kuhn, T. S. (1970). The structure of Scientific Revolutions. In *International Encyclopedia of Unified Science* (Issue 2). The University of Chicago Press.

Lakatos, I. (1978). Science and Pseudoscience. *Philosophical Papers*, *1*, 1–7.

Manson, S. M. (2001). Simplifying complexity: a review of complexity theory. *Geoforum*, 405–414.

Melamed, Y. Y., & Lin, M. (2021). Principle of Sufficient Reason. In *The Stanford Encyclopedia of Philosophy* (Summer 202). https://plato.stanford.edu/archives/sum2021/entries/sufficient-reason/

Milas, G. (2009). Istraživačke metode u psihologiji i drugim društvenim znanostima [Research methods in psychology and other social sciences]. Naklada Slap.

Nagel, E. (1961). *The Structure of Science: Problems in the Logic of Scientific Explanation*. Harcourt, Brace & World Inc.

Physics and Beyond: "God does not play dice", What did Einstein mean? (2021). https://www.stmarys.ac.uk/news/2014/09/physics-beyond-god-play-dice-einstein-mean/

Popper, K. (1963). Science as Falsification. *Conjectures and Refutations*, *1*, 33–39.

Reichenbach, H. (1956). *The Direction of Time*. University of California Press.

2020 *Science*. (2020). Wikipedia. https://en.wikipedia.org/wiki/Science

Skinner, B. F. (1990). Can Psychology Be a Science of Mind? *American Psychologist*, *45*(11), 1206–1210. 10.1037/0003-066X.45.11.1206

Sundet, J. M., Barlaug, D. G., & Torjussen, T. M. (2004). The End of the Flynn Effect? A Study of Secular Trends in Mean Intelligence Test Scores of Norwegian Conscripts During Half a Century. *Intelligence*, *32*, 349–362. 10.1016/j.intell.2004.06.004

Teasdale, T. W., & Owen, D. R. (2005). A Long-term Rise And Recent Decline In Intelligence Test Performance: The Flynn Effect In Reverse. *Personality and Individual Differences*, *39*, 837–843. 10.1016/j.paid.2005.01.029

Walker, T. C. (2010). The Perils of Paradigm Mentalities: Revisiting Kuhn, Lakatos, and Popper. *Perspectives on Politics*, *8*(2), 433–451. 10.1017/S1537592710001180

Wallace, A. F. C. (1959). Cultural Determinants Of Response To Hallucinatory Experience. *A.M.A. Archives of General Psychiatry*, *1*(1), 58–69. 10.1001/archpsyc.1959.03590010074009

Watson, J. (1913). Psychology as the Behaviorist Views it. *Psychological Review*, *20*, 158–177. http://psychclassics.yorku.ca/Watson/views.htm

Wickstrom, G., & Bendix, T. (2000). The "Hawthorne effect"-what did the original Hawthorne studies actually show? In *Stand J Work Environ Health* (*Vol. 26*, Issue 4).

Wolfram, S. (1984). Universaility and Complexity in Cellular Automata. *Physica*, *10D*, 1–35.

Woodward, J. (1980). Developmental Explanation. *Synthese, 44*(3), 443–466. 10.1007/BF00413471

Woodward, J., & Ross, L. (2021). Scientific Explanation. In E. N. Zalta (Ed.), *The Stanford Encyclopedia of Philosophy (Summer 2021 Edition).*

Wray, K. B. (2011). Kuhn and the Discovery of Paradigms. *Philosophy of the Social Sciences, 41*(3), 380–397. 10.1177/0048393109359778

2   Basic concepts of statistics

We will start this introductory story with presenting the basic concepts statistics is based on. While the previous chapter discussed philosophical bases of science in general and statistics in particular to help the reader understand the place statistics has in the world of science, this chapter starts introducing the reader to statistics itself by presenting and describing key concepts statistics is based on.

2.1 Random events

In statistics theory, a random event is an event outcomes of which are unpredictable in principle, an event that can produce different outcomes under the same set of conditions. Random events are the building blocks of a stochastic world (see chapter 1.2.) as they are the component that makes the stochastic nature of the world possible by creating situations where same sets of conditions lead to different consequences. Although a random event is defined as an event the outcome of which is essentially unpredictable, it is generally considered that probabilities i.e. frequencies with which various outcomes occur can be observed and these frequencies then used to predict the frequencies with which these same outcomes will occur in the future. That said and remembering the discussion from chapter 1.2 on how we currently do not know for certain whether the nature of our universe is deterministic or stochastic, it should be noted that random events are at this point largely a theoretical concept, a vessel of the statistics theory needed to develop and apply statistical procedures. At this point in time, there is no known way to produce an event for which we would be certain that it is random, nor is there any known category of events for which it can be said with certainty that they are random. Yes, there are many classes of phenomena and classes of events, that we treat as random events because we do not know how to predict them accurately, but there are no events for which we can say with certainty that they will not become predictable in the future.

So, if statistics is based on random events, but we cannot produce random events nor do we recognize any random events, how does statistics work? It works by applying the best possible approximation – the so-called pseudorandom events. **Pseudorandom events** are events that are not truly random or likely not truly random, but such that we cannot predict them, or intentionally refrain from predicting them (as that would be "cheating" and negate the very purpose of their generation) and that are not causally connected to the purpose for which we need them. The most commonly used type of pseudorandom events are **pseudorandom numbers** obtained through the use of devices that are collectively called **pseudorandom number generators**. These devices

DOI: 10.4324/9781003107712-2

range from very simple ones like dice (cubes with numbered sides used in games of chance) or pulling numbered balls out of a dish to very complex ones based on the decay of radioactive atoms, fluctuations of natural radio wave emissions or other physical phenomena, or on complex computational procedures (e.g. Akhshani et al., 2014; Desai et al., 2011; Liu et al., 2021). Pseudorandom number generators, generally consist of two components – a seed i.e. initial values that are input into the generator and the algorithm or the mathematical function that transforms these initial values into the end number that conforms to requirements of the generator (such as being within a certain range). This fact that a pseudorandom number generator uses an algorithm or a set of mathematical operations to generate its final values, reminds us again that pseudorandom number generators are not true random number generators. It is of course possible that the seed be a random event, but so far there is no known way of creating input values that would be truly and certainly random. Using natural events that we cannot predict or that are hard to predict might make these initial values unpredictable to us, but that is still not a proof that they are unpredictable in principle. Given this, the most important property of pseudorandom number generators is the statistical properties of their outputs and the extent to which they conform to expectations of the statistical theory about a random event and about how truly random numbers would look like. However, as similar to theoretical expectations of how a random event will look like as they might be, one should not equate them with truly random events or with the generation of truly random numbers that would be possible if we had such an event. As the famous mathematician and physicists John Von Neumann is attributed saying "Anyone who considers arithmetical methods of producing random digits is, of course, in a state of sin".

2.2 Probability

Another key concept based on the premise of the existence of random events is probability. **Probability** is usually defined as the **likelihood that a certain outcome of a random event will occur**. It is usually expressed as a number in the range between 0 and 1, denoting **the proportion of the total number of outcomes in which the event occurs**. A probability of 0 means that that particular outcome never occurs as the outcome of the random event, while a probability of 1 means that that particular outcome is always the outcome of the random event in question.

Formulated in this way, probability is presented as an expectation on how future events will unfold. However, probability is calculated based on past events and the expectations that future will be the same as the past was. **Probability is calculated by making a large number of observations of an event we view as (sufficiently) random and counting the number of times different outcomes occurred.** Then we divide the number of each particular outcome by the total number of observed outcomes (of all types) and declare that this proportion is the probability of that particular outcome. Presented mathematically, it looks like this:

Probability of outcome A = number of times outcome A was observed

/total number of observed outcomes

A key take here is that the total number of observed outcomes needs to be large in order for it to be possible for probability to be assessed like this. This has to do with a mathematical theorem called the **Law of large numbers** that states that the ratios of outcomes of an observed random event will become closer to their true probabilities as the number of observed events (referred to as trials) increases. When the number of observations is small it is much more likely that our outcomes deviate more from their true probabilities. For example, when throwing dice, it is much easier to obtain 6 in 2 throws in a row than in 10 throws in a row. If we threw dice a million times, it would be practically impossible that we get a 6 every time in those million times. This of course, assuming that the dice has the same probability of landing on each of its sides (each of which is marked with a different number from 1 to 6).

Now, as we discussed in previous chapters, assessing probability like this implies **two "leaps of faith"**:

- that **the event we are observing is truly random,** while it is either not or it is not certain that it is random. However, we can refer to past observations to note whether it has behaved in the past in a way that we would expect a random event to behave. If this is the case, we can declare that it is sufficiently similar in behavior to a random event to be treated as such.
- that **the observed event or class of events will behave in the future in the same way it has behaved in the past**, meaning that the likelihoods of occurrence of particular outcomes, i.e. their probabilities, will remain the same. Now, people who apply statistics in practice have a well-known saying that past trends and success of past statistical predictions are often poor indicators of future trends and the success of future predictions. Belief that phenomena will behave in the future in the same way they behaved in the past, without knowing enough about their nature to support this notion is at least a risky proposition. However, when nothing better is available, this approach becomes the best available option (see statistical explanations in the previous chapter).

The study of probability is the topic of a branch of mathematics called probability theory.

2.3 Entity

Statistical observations and statistical calculations are usually done on certain properties of certain objects. These objects can be of any possible nature and they widely differ in various scientific areas in which statistics is applied. In social sciences, these objects may be persons, or groups of people or organizations. In biology they may be organisms, plants, animals. Or they may be paintings, or pieces of equipment, or subatomic particles or parts of objects etc. And they are usually studied in sets, as that is how statistics works.

Whenever we wish to refer to an individual member of a statistical set without specifying its nature, we call it an entity. So, **an entity is an individual member of a statistical set regardless of the nature of that member.** It can be a person, an animal, an object, a piece of an object or anything. As long as it is included in a statistical set, we can refer to it as an entity.

2.4 Variables and constants

As said previously, statistical observations and statistical calculations are done on certain properties/features of certain objects. These properties may differ in value for different entities or they may have equal values for all entities. For example, if our property is the hair color of a person, its values may be various existing hair colors such as black, blond, red, blue, orange etc. If our property is the height of a person, then values of that property can be various heights a person can have such as 180 cm, 165 cm, 182 cm, 190 cm etc. These properties can be such that all observed entities have the same value on them or such that different entities have different values of the property. If all observed entities have the same value of the property, that property is called a **constant** for that group of entities. An example of a constant would be if we counted the number of heads a person has. Given that every person has exactly one head, the number of heads per person would be an example of a constant[1]. If a property/feature is such that different entities have different values of that property, than such property is called a **variable.** Previously mentioned height and hair color are examples of variables. Certain personality dimensions such as, for example, extraversion can also be an example of a variable as people can be extraverted to different degrees and this can be assessed through psychological testing. Price of a certain commodity can also be a variable if we, for example, take different days (on which the price is registered) as entities. And so on. What is important to note is **that statistical procedures are primarily conducted on variables** while there is little use for statistics when we are dealing with constants. And generally, the more values of entities on a variable differ, the more they vary, the more we can do with the use of statistics. Practically all existing statistical procedures, and definitely all the procedures that are discussed in this book are meant to be used on variables and not used on constants. They produce meaningful and useful results only when used on variables! When used on constants, there is usually not much more we can learn about it except the fact that we are dealing with a constant. Therefore, whenever considering a property of the entities we are observing, one of the first things to note is whether we are dealing with a constant or a variable.

One more thing that experience in teaching introductory statistics has taught us to emphasize when introducing the concept of a variable to students is to underline the difference between a variable per se and a value of that variable, as this is something students often confuse in the start. A variable is a property or a feature of an entity that can change and assume different values, and variable values are possible values a variable can have. For example, we can talk about a variable called weight (of a person) and its values are different weights in kilograms (or any other unit) a person can have. Similarly, if a group of students took a test that was graded from A to F, the variable would be "grade", while variable values would be A, B, C, D and F (the possible grades).

2.5 Organization of data, matrix, vector

Now that we know that statistical procedures are done on values of variables that were observed on groups of entities, we come to the question of how to organize the data to make it possible to do calculations with it. Typical statistical procedures in modern science usually involve the use of rather large quantities of data, and the use of software tools to perform the calculations. Actually, there is a very limited number of situations today where practically useful statistical calculations can reasonably be done without the use of computers and statistical software. Due to this, the question of how to organize the data becomes a question of prime importance.

The most typical way observational data, i.e. data about values of a group of entities on a number of variables, is organized for performing statistical calculations is in a matrix (data matrix). A **matrix is simply a table with a certain number of columns and a certain number of rows**. It is named based on the number of rows and columns. For example, if a matrix has 100 rows and 200 columns it is named 100×200 matrix (or 200×100 depending on the naming rules of the entity doing the naming). Most typically, entities are represented by rows and variables by columns. If we follow a row, we can read values of a certain entity on all the variables. Following a column shows values of all entities on that specific variable represented by the column. There are, of course, instances where data is presented differently, but the default presentation used in practically all statistical software is this – rows for entities, columns for variables (Tables 2.1 and 2.2).

Table 2.1 An example of a data matrix. Entities are people who have taken a personality test, variables are their names and their scores on various personality inventory components (fictional data)

Name	Neuroticism N	Extraversion E	Agreeableness A	Openness to experience O	Conscientiousness C
Becky	45	55	55	65	70
Anita	22	67	50	72	62
Vladislava	37	25	45	65	45
Careen	55	32	65	42	35
Esmaeel	25	25	32	28	60
John	52	12	42	35	28
Hamza	50	70	48	47	32
Vladimir	38	65	51	59	65
Emmet	40	25	65	61	65
Mark	25	30	22	45	50
Ellen	32	45	45	68	65
Peter	35	55	65	42	58

Table 2.2 Another example of a data matrix. Entities are stocks of various companies and variables are their stock market ticker symbols and various financial details about the company as publicly reported at the time of writing (data are for illustration purposes only and are not accurate)

Company/ticker	Price at last close	EPS	P/E ratio	P/S ratio	P/B ratio	Beta
AMD	107.56	2.8	38.4	9.78	18.5	2.01
DADA	173.73	8.3	20.9	3.97	3.1	0.81
MSFT	292.52	8.1	36.3	13.08	15.5	0.78
ARCB	68.37	5.1	13.3	.52	1.9	1.78
ENPH	163.48	1.27	128.9	19.87	36.1	1.17
HLX	3.63	.06	60.3	.81	.3	3.39
PLAN	59	−1.18	−50.3	18.03	32.5	
QCOM	144.41	8.01	18	5	19.9	1.32
CLVT	22.61	−0.6	−37.65	8.99	1.4	
FSLY	40.96	−1.58	−26	14.78	4.6	
EXEL	19	.29	66	5.18	2.9	1.05
W	298.38	3.06	97.7	2.09	−20	3.09
IQ	8.75	−1.09	−8.05	1.47	5.5	.8
JD	64.26	5.04	12.7	.82	3.4	.76

Abbreviations: EPS – earnings per share; P/E – price to earnings; P/S – price to sales; P/B – price to book value; Beta – Beta coefficient of stock price volatility.

Table 2.3 An example of a data matrix format used for presenting results of statistical calculations (Tošić Radev & Hedrih, 2017). Both columns and rows represent variables and the number at the intersection of a row and a column is the correlation (a measure of association/a statistical indicator of the degree of joint variation) of the variable represented by the row and the one represented by the column

Correlation coefficients between the Multidimensional jealousy scale and assessed external variables

	Scale	Cognitive jealousy	Emotional jealousy	Behavioral jealousy	Overall jealousy
	Global Self-Esteem	$-.32^*$	$-.11^*$	$-.13^*$	$-.25^*$
BFI	Neuroticism	$.32^*$	$.27^*$	$.27^*$	$.36^*$
	Extraversion	$-.12^*$	$.01$	$.03$	$-.05$
	Openness to Experience	$-.12^*$	$-.14^*$	$-.14^*$	$-.16^*$
	Agreableness	$-.10^*$	$-.07$	$-.13^*$	$-.12^*$
	Consciousness	$-.20^*$	$-.04$	$-.09^*$	$-.15^*$
LAS	Eros	$-.17^*$	$.05$	$.05$	$-.05$
	Ludus	$.16^*$	$-.07$	$-.03$	$.04$
	Storge	$-.02$	$-.05$	$.01$	$-.02$
	Pragma	$.07$	$.06$	$.13^*$	$.10$
	Mania	$.39^*$	$.37^*$	$.44^*$	$.50^*$
	Agape	$-.09^*$	$.03$	$.11^*$	$.01$

All correlation coefficients higher than .09 are statistically significant at least at the .05 level.

Also, aside from presenting observational or measurement data, the matrix form is also used for presenting results of statistical calculations and in these cases the methods of presentation can vary and depend on the contents that are to be presented in the matrix and the preferences of the person making the presentation. For example, Table 2.3. presents a data matrix presentation of measures of associations between variables used in a study. Both columns and rows represent variables and the numbers are coefficients indicating the degree of joint variation – correlation between variables, table taken from Tošić Radev & Hedrih (2017).

A matrix that has the same number of columns and rows is called a **square matrix**. Square matrices are typically a method for presenting relations between variables where the same variables are represented by both rows and columns and matrix cells are filled with whatever measure of relations between them is to be presented. A matrix where the number of rows and columns differs is called a **rectangular matrix.**

There are also certain specific types of matrices that are an important component of various statistical procedures, usually as a point of reference for comparison, that have specific names. Such matrices include:

- **diagonal matrix** is a square matrix in which cells along the main diagonal contain various numbers, while all other cells in the matrix contain the number 0. A diagonal matrix is usually used as a reference point when describing relationships between variables to denote a situation when the measure of a relationship between different variables is 0, while the variable may have a non-zero relationship with itself or a corresponding variable (presented in the corresponding row/column). Typically, in this type of matrices, the same variables are both represented by rows and columns or that there are two sets of corresponding variables, one represented by rows and the other by columns. A diagonal matrix can also be a intermediary product of certain statistical calculation procedures.

• **identity matrix** – is a variant of the diagonal matrix, it is a square matrix where all cells along the main diagonal contain the number 1, while all the other cells contain the number 0. Similar to a diagonal matrix, an identity matrix is also used as a reference point when describing relationships between variables to denote a situation where the measure of relationship between different variables is 0, while the variable is related to itself (hence 0's on the diagonal) is 1 (Tables 2.4 and 2.5).

A matrix that consists only of a single column (regardless of the number of rows) or a single row (regardless of the number of columns) is called **a vector.** A vector is usually a representation of all values of a single variable across all entities or a representation of values of a single entity on all variables, but it can also be a representation of relations of a single variable with all other variables in a dataset and can also be used in various other representations. Formally it is a matrix with a size of $1 \times n$ or $n \times 1$, where n is any natural number. In so-called informational geometry, which is a specific type of application of geometry in statistics, it can be used as a representation of coordinates of a single entity in the considered statistical space by containing values of the entity on all variables, which are essentially its coordinates in the statistical space (Table 2.6).

Table 2.4 A diagonal matrix, an example. All elements are 0s, except those on the main diagonal that can be any number (fictional data)

.22	0	0	0	0	0	0	0
0	45	0	0	0	0	0	0
0	0	11	0	0	0	0	0
0	0	0	22	0	0	0	0
0	0	0	0	−25	0	0	0
0	0	0	0	0	35	0	0
0	0	0	0	0	0	−2	0
0	0	0	0	0	0	0	55

Table 2.5 An identity matrix, an example. All elements are 0s, except those on the main diagonal that are 1s. Similar to a diagonal matrix, an identity matrix is used as a reference point when describing relationships between variables to denote a situation where the measure of relationship between different variables is 0, while the variable is related to itself (hence 0's on the diagonal) is 1

1	0	0	0	0	0	0	0
0	1	0	0	0	0	0	0
0	0	1	0	0	0	0	0
0	0	0	1	0	0	0	0
0	0	0	0	1	0	0	0
0	0	0	0	0	1	0	0
0	0	0	0	0	0	1	0
0	0	0	0	0	0	0	1

Table 2.6 Two examples of vectors. Vectors are matrices consisting of only a single row or a single column. The first example is a vector containing Vladislava's data on all variables. The second example is a vector containing values of all study participants on the personality trait Openness to experience

Name	Neuroticism N	Extraversion E	Agreeableness A	Openness to experience O	Conscientiousness C
Becky	45	55	55	65	70
Anita	22	67	50	72	62
Vladislava	37	25	45	65	45
Careen	55	32	65	42	35
Esmaeel	25	25	32	28	60
John	52	12	42	35	28
Hamza	50	70	48	47	32
Vladimir	38	65	51	59	65
Emmet	40	25	65	61	65
Mark	25	30	22	45	50
Ellen	32	45	45	68	65
Peter	35	55	65	42	58

2.6 Population and sample, parameters and statistics

Statistical procedures are generally conducted in order to infer or discover something about a large group or a category of entities. A large group of entities that we wish to study is called a population. For example, a population may be all people living in a certain country or in a certain area, or all humans everywhere, or all animals of a certain species in an area or anywhere. But populations can also be inanimate objects like all the water in a river, all the water in a specific lake or a sea, all paintings of a certain painting style, all organs of a certain type in certain types of organisms, all prices of a certain commodity on different days or at different times, all prices of different commodities, all measurements of a certain property of objects etc. Populations can be very different in their nature, but for a certain group to be considered a population in the statistical meaning of the word it needs to be precisely defined so that it is perfectly clear when considering any entity whether that entity belongs to that population or not. Populations can also be fixed or limited in nature, or they can be (practically) unlimited. For example, if we define our population as students of certain specific ongoing class of a certain school, that is a clearly limited population as we can clearly make a list of all members of that population and it will likely not change until it is disbanded i.e. until the school program is finished and the class is disbanded. There might be new students in a class, or some dropping out, but at any given point, we can know which students exactly belong to that class and this does not change or changes very little. In contrast to this, we can also specify that our population of interest might be "all humans" or "all humans living in a certain territory". If we take "all humans" to be our population of interest, then this includes all people who ever lived and who are yet to be born and this is a population of which we clearly cannot create a list, because even if we had a list of all people who ever lived, we have no way of knowing who exactly will be born in the future. However, as we will see later, a population defined like that is too broad to really be studied using statistical methods, so we would practically have to narrow it down for research purposes. We could, for example, define our population as "all people living in a certain area around the time of the study". While this is narrower, making a definite list of this population would also be practically impossible, because making a list takes time, and in the time it takes to make the list, the contents of the list change as some people die, some people move away, some move in, some are born etc. And till the time we manage to compile the list it is already not completely accurate. It is also likely that, barring some sci-fi type device, it would be practically quite hard or impossible to track all people living in any large area, because not all are registered and we would practically have trouble finding everyone. The thing becomes even more complicated if our population is "water in a specific river" or "insects of a certain species living in a certain large area" etc. However, the good news is that we can organize research even if we cannot list all the members of a population as long as the population is defined with sufficient precision to allow us to clearly differentiate between entities that belong to the considered population and those that do not. The distinction between limited populations and unlimited populations comes into play when making statistical inferences about them based on studying a part of the population.

In an ideal situation, when we want to study some property of the population i.e. some variables in the population, we would take measures or assessments of these variables on all members of the population and make our conclusions from the results. This is practically possible when we are studying populations that are limited and also small. If a population is not small, taking assessments/measurements from all members of the population quickly becomes impractical and expensive and if the population is not limited in numbers, then it becomes literally impossible.

However, taking measurements from the whole population is most often not necessary. If our research aim is to obtain findings about properties of the population as a group and if we are not interested in the individual values of each member, studying every entity in the population becomes unnecessary. For example, if we want to find out what the typical price of tomatoes is on a market, we do not really need to ask every tomato seller for the price of his/her tomatoes. We can record the prices of a certain number of randomly picked sellers and gain a pretty good idea of what the price range might be on that market. If we want to discover the content of, for example, some chemical compound in a lake, we do not need to count every molecule of that compound in the whole lake. Usually taking a full bottle of water from several different points in the lake will give us a pretty good idea of the general content. If we want to assess the general level of knowledge of a certain foreign language in the population, we can assess it in a randomly selected group of members of that population and that would give us a pretty good idea of the general knowledge level of the population. Speaking in statistical terminology – we would study a sample to make inferences about the population.

A part of the population that is selected for study is called **a sample.** The general idea behind studying a sample is that by studying a sample we will learn what we wish to learn about the population because the sample is sufficiently similar to the population that we can make inferences about the state of affairs in the population based on what we learn from studying the sample. Yes, properties of the sample might sometimes not be exactly the same as the population, but we can also expect that most of the time they will not be too different. If we take a sample that is large enough and do what we can to avoid intentionally selecting a sample that is different from the population, the sample will likely be sufficiently similar to the population to allow for many useful inferences about the population.

What should a good sample be like? Ideally, all the properties of the sample should be the same as the properties of the population as a group, with the only difference being that the sample is sufficiently small to be studied (unlike, most of the time, the population). A sample that has all properties equal to those of the population (except size i.e. the number of entities in it) is called a **representative sample.** When we conduct a study that aims to discover something about the properties of the population, we would ideally like the sample we conduct our study on to be representative. But can we know whether the sample we collected indeed is representative of the population? This is a bit of a chicken and egg question – to know for sure whether our sample is representative, we would need to compare it to the population. However, to compare it to the population, we would need to have data for the population. But if we had data for the population, we would not be needing a sample in the first place. This means that **in real life situations, we can never know for sure whether our sample is representative for the population it is meant to represent**. Yes, most of the time there will be heuristics that may help us notice if our sample is terribly, obviously, off from the population. For example, if we collected a sample of people who are all young and we know that the population we are interested in has a much larger age variability, it would be obvious that such a sample is off. Or if we collected a sample that solely consists of males, but we notice through casual observation that our population is gender diverse and does not consist solely of males. However, in cases when the difference between the sample and the population is not so extreme, obvious or pronounced, there would be no way to tell whether our sample is representative. This means that, although there might be cases when we can tell that our sample is not representative, **there is no way to be positively sure that a sample is representative.** Yes, there are various ways in which researchers can try to maximize the chance that a sample is representative, such as using specific techniques for collecting a sample that have typically resulted in relatively representative samples in the past, or making assessments based on various statistical properties of the sample and comparing them to those same data from previously collected samples, but none of these guarantee that the sample collected in a certain (practically possible!) way or that passed a certain statistical assessment is representative. Nothing short of directly comparing the sample to the population can tell us for sure whether the sample is representative and as said earlier, this comparison is impossible in most practical situations.

This means that no matter how we select a sample, there is always a chance that it will be more or less different from the population it is meant to represent. And if we accept that the possibility of this difference cannot be avoided, this means that it must be, in one way or another, incorporated into the way we assess the properties of the population based on the sample. For this reason, statisticians created **the concepts of statistics and parameters**. When we assess certain properties of a sample, for example when we calculate average values of a variable on a sample, or when we count the number of entities in a sample that have a certain value on a certain variable, the result of these procedures is called a statistic. When we do the same on a population, the result is called a parameter. However, as said before, most of the time, we cannot measure values of a variable on a population directly, so we have to infer about those values based on the values of the sample. In other words, any statistical indicator calculated from sample data is called a **statistic**. A statistical indicator of a population is called a **parameter.** Typically, we directly calculate statistics from the sample, while we make inferences about parameters based on the values of statistics. We typically cannot calculate parameters directly from data, as that would require having data on all members of the population. Methodology of making such inferences can be quite complex and it is actually the topic of a branch of statistics called **inferential statistics.** Basics of inferential statistics will be presented in the second half of this book (Figure 2.1).

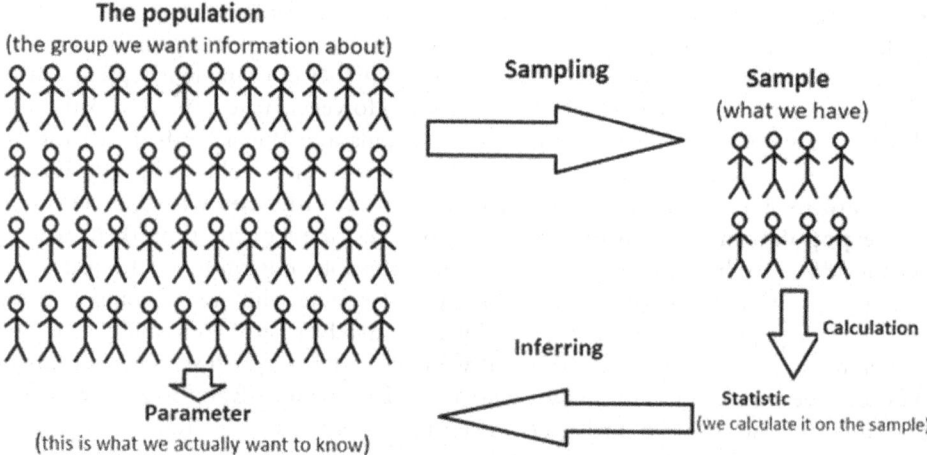

Figure 2.1 The population and the sample, how things work in statistics.

2.7 Sampling

The procedure of selecting a sample of entities to be used in a study from a population is called sampling. As mentioned before, for a study on a sample to be successful and provide useful information about the population, the sample needs to have certain characteristics, most common of which is that it be as representative as possible of the population we want to make inferences about[2]. To achieve this goal, there exist different techniques for building a sample – sampling techniques that differ in the way entities are selected for the sample, the ease of creating such a sample and, most often, in the likelihood that a sample created in that way will be representative. We can differentiate between general approaches to sampling and specific sampling techniques.

When considering general approaches to sampling we should first underline the difference between sampling without replacement and sampling with replacement.

Sampling without replacement is a sampling procedure in which an entity selected for inclusion in the sample is moved from the population to the sample and cannot be selected for inclusion in the sample again. That means that each entity from the population can be included in a single sample exactly once – there cannot be multiple copies of the same entity in the sample. The main effect of this is the appearance of, so-called, rising probabilities. If we select entities to the sample using random selection, where each entity in the population has the same probability of being selected for the sample, as sampling progresses the probability of each entity in the population for being selected for the sample increases as the number of entities available for selection decreases. It decreases as some entities are already included in the sample and are no longer available for selection. For example, if we have a population of 1000 entities and we select them randomly so that each one has the same chance of being included in the sample, when selecting the first entity for the sample, each entity in the population would have a chance of 1/1000 to be selected for the sample in that step (as we are selecting one out of 1000, and each entity has the same chance, the probability is than 1/1000 for each entity). But already in the next step, we are selecting 1 out of 999, making the probability 1/999. In the third step it would be 1 out of 998, then 1 out of 997 and so on. This might be especially important if the population size is limited and the size of the planned sample is a substantial proportion of the population size. This effect is practically irrelevant if population is unlimited or if the sample size is negligible compared to the size of the population. This method also prevents us from making samples that are larger than the population.

Sampling with replacement is a sampling procedure in which, once an entity is selected for inclusion in the sample (sampled), that entity is "returned" back to the population so that it can be sampled again. In this way it is possible for the same entity to be sampled multiple times for the same sample. The result of this is that the same entity from the population may, in a way, be represented as several entities in the sample. While this might seem like an awkward idea at first glance, there are various quite useful effects of this type of sampling. The first and the most obvious one is that, in contrast to sampling without replacement, when doing random sampling with all entities in the population having equal probabilities of being selected, these probabilities do not increase as the sampling progresses, but remain constant. When doing random sampling without replacement from, for example, a population of 1000 entities and with equal probabilities of being sampled for all entities, each entity will have a 1/1000 chance to be selected both in the first step and in all subsequent steps. This is due to the fact that entities that are selected into the sample in one step are again available for selection in subsequent steps, so in each step we are selecting from the full population size. Another important effect is that we can make any sample size from any population size with this procedure. This means that we could make samples that are larger than the population and that we could have a large, even infinite number of samples from the same limited population and the composition of these samples could differ more than it would be possible if they were sampled without replacement. For example, if we made 100 samples that are the same size as the population using sampling without replacement, they would necessarily be all the same, as they would consist of exactly the same entities. However, if we did the same thing using sampling without replacement, these samples could differ as there would likely be multiple copies of same entities in each sample, but of different ones and in different numbers in different samples, while some entities would not be included in different samples, allowing for these samples to differ between each other.

Another important general property of sampling techniques is whether the method for selecting entities is probabilistic or nonprobabilistic. In **probabilistic sampling** procedures, selection of entities is based on a stochastic process (or, in practice, an available approximation of such a process, see chapter 1.), with each entity in the population being assigned a certain probability of being selected into the sample. These probabilities may be different for different entities or can be the equal, but for the sampling procedure to be considered probabilistic, sampling must be performed through chance and probabilities. In contrast, **nonprobabilistic sampling** procedures use a variety of sampling procedures in which selection is not based on chance or probability.

At this point it is also important to point to the concept of resampling. **Resampling** refers to a number of procedures where a new sample or samples are created from an existing sample, typically in order to simulate what would happen if additional samples were drawn from the same population. As we are aware that no matter how we create a sample its properties may more or less differ from the population, resampling can be a convenient way to assess the likely magnitude of these differences and thus make more precise inferences about what might be expected from future studies of the same topic using different samples. While there are many different methods of resampling, those most commonly seen in scientific research and statistical software include bootstrapping, jackknifing and cross validation.

Bootstrapping is the name of a set of procedures where a (typically large) number of new samples is sampled with replacement from the one sample at hand (e.g. Good, 2006). Essentially, the existing sample is treated as if it were the population and then a number of new samples are sampled from entities in it. This is usually done by using random sampling (see later) and because it is done through sampling with replacement there is no limit in the number of new samples that can be sampled or in their size. Bootstrapping is most commonly used in procedures for making inferences about parameters (i.e. values of statistical indicators in the population) based on the statistics calculated from the sample. In a way, bootstrapping can be taken to simulate what would happen if we sampled a large number of samples from the same population, with the only difference being that here we are not sampling from the population but from a sample taken from that population. This will be discussed in more detail in the part of this book about inferential statistics. Bootstrapping procedure was first proposed by Efron (1979) and further developed by various other researchers. At the moment this book is written, it is becoming a commonplace procedure of inferential statistics found more and more in mainstream statistical software.

Jackknifing is a procedure in which multiple samples are created by excluding a certain number of entities from each sample. This is the same as sampling without replacement a number of samples from an original sample. The procedure was proposed in 1949 by Quenouille and further developed by Tukey in the 1950s (Miller, 1974). The idea behind it is to assess how much sample statistics change when we exclude a part of a sample and thus, in a way, make an estimation of what would be obtained if another sample was taken from the same population and sample statistics calculated on it.

Cross validation is a procedure in which a sample is divided into two parts. Statistical calculations are then performed and inferences made on the first part of the sample and it is then examined if these same inferences hold on the second part of the sample. It is yet another way of estimating how valid any generalizations from the sample to the population or, more practically, other samples from that population are.

2.8 Types of samples

Now we will go through specific types of samples i.e. specific sampling procedures and discuss each of them. As said earlier, types of samples differ in the ways entities are selected into the sample and, through that, they differ in how hard it is to collect such a sample and how likely it is that the properties of the sample will differ significantly from those of the population. This is not an exclusive list of all possible types of samples, because such a list would be unlimited as any particular way in which someone creates a sample can be considered a sampling technique in itself. These are just some of the most common sampling procedures:

Simple random sample is created by making a list of all entities in the population to be studied and then using a random number generator to select those that will be included in the sample. The random number generation procedure is created in such a way that all entities in the population have an equal probability of being selected into the sample. Simple random sample is what people usually refer to when they speak about random samples. It can be said that this type of sample is a sort of a "model sample" in statistics, being the basis of most statistical theorems and inference models and with all the other types of sampling seen as better or worse approximations of this type of sampling. However, in most real-world situations, this type of sample is impossible to produce in practice. First, a simple random sample requires that we be able to generate random numbers. This is, of course, impossible, so pseudorandom numbers are used instead. Also, it requires the population to be studied to be limited and that we have a complete list of all entities in the population. This makes a simple random sample not an option in all cases where we have unlimited populations or when we do not have a list of entities comprising the population. Finally, this type of sample requires that we have the power to force every entity that we select through random number generation to become a part of the sample. When these entities are people, such as in the case of studies in social sciences, forcing everyone selected into the sample is, of course, impossible, as people may and do refuse to participate in studies. If entities are not people but natural objects, such as molecules of water the population may not be limited and even when it is limited, such as, for example, if our population consisted of trees in a particular forest, it is often impractical to make lists of all entities (e.g. all trees in a large forest) just for the purposes of sampling. So, although simple random sample is a sort of a "model sample" of statistics, it can rarely, if ever, be used in practice. However, this is not really a great problem as there are other sampling techniques that are much easier to perform in practice, but that typically result in samples that are not much worse than random samples.

Convenient sample is created by simply including into the sample those entities that can most conveniently be sampled, that are at hand, those that are most easily available to the researchers. In contrast to the simple random sampling, convenient sampling is usually considered the worst sampling technique. The main problem with convenient sampling is that the quality of the resulting sample i.e. how representative of the population it is, can vary widely. Sometimes convenient sampling can result in samples that are quite similar to the population, but other times it can result in samples that miss the population parameters by a huge margin. And it might often be hard or impossible to tell which of these will be the case when looking at a specific sample. That said, in various social sciences, convenient sampling can sometimes be aided with the personal expertise of the researcher to improve the match between the sample and the population. For example, researchers will often have an idea about what general properties the population under study has, and might than be able to tell which groups or parts of the population might look like typical representatives of the population and which might not and also have a good idea where they can easily take a sample from, without that sample being too different from the population. There is, of course, also the danger of this being done with the opposite intent – a researcher with an unscientific (political, ideological, pseudoscientific…) agenda, one that does not really want to do research but is purely interested in obtaining the results he/she prefers (for the purpose of misleading the academic audience) can select a convenient sample for which he/she knows that it differs from the population but selects it nonetheless because it would provide the results he/she wants. So, in most aspects the use and quality of convenient sampling depends on the expertise of the researcher, but also his/her scientific honesty and integrity. That said, **convenience sampling is by far the cheapest, easiest and the least demanding sampling technique**, one that is an option whenever resources of the researcher are tight. That is the main reason why convenience sampling is the method used in the vast majority of studies in the area of social sciences. However, it is also the reason why it is less common in the most influential studies than it is in studies in general.

Stratified sample is created by dividing the population into strata i.e. into sub-populations based on a certain trait and then selecting entities from each stratum using some of the other sampling methods (random, convenient sampling etc.) to be included in the sample. The idea is that, if the population consists of certain distinct subgroups that are important for the topic of the study, stratified sampling makes certain that all these subgroups are adequately represented in the sample. Ideally, definitions of strata will follow some obvious separation into subgroups, so that it is easy to conduct sampling within each stratum (example might be different grades in a school or different municipalities within a state). Stratified samples can be made to be either **proportional**, when the share of entities from each subpopulation is in proportion with the share of that stratum in the population or **disproportional**, when the share of entities from each stratum is not proportional to the share of the stratum in the population. Disproportional sampling can be useful when we want to have more participants from certain strata, for example in situations when there are strata that are so small that, if the sampling were proportional, there would be too few entities from those strata for adequate study. It should also be noted that, while disproportional stratified sampling is a perfectly legitimate sampling technique, it is important that all researchers involved in the study, including data collection and interpretation, maintain throughout their work with the sample the awareness that they are working with a disproportional sample and refrain from making inferences about the population as a whole in a way they would do with a proportional sample. When working with a disproportional sample, the fact that it is a disproportional sample needs to be taken into account and compensated for (for example by assigning so-called "weights" to different strata) before making inferences about the population based on that sample.

Quota sample uses a system of quotas to create the sample. Quotas are numbers of entities with certain values of important variables that are to be included in the sample. They are defined based on the proportions of entities in the populations with those variable values and then the sample is created so that proportions of entities with those variable values are the same as in the population. The idea behind this sample is that if a sample is created so that it resembles the population with regard to the distribution of values of certain key variables, it is more likely that it will resemble the population with regard to all the other variables, including those that are to be studied on the sample. Of course, for a quota sample to be possible, distributions of values of the variables that will be used for creating quotas in the population must be known. When quotas are defined, any available entities that fill the requirements of each quota are selected. For example, if we know that our population of interest consists of 15% of university students and of 85% of people who are not students, we can create a quota sample of 100 people by defining that it should consist of 15 students and 85 non students. These numbers of students and nonstudents to be included in the sample are quotas for that sample. This is the simplest example where only one variable with two categories is used to create quotas. Typical quota samples however, include quotas based on multiple variables. In this case, these quotas can be tied or untied. **Untied quotas** are quotas that are defined for each of the quota variables separately, without crossing them. For example, if we made quotas based on whether a person is a university student or not (one variable) and on whether that person is above or below 40 years of age, untied quotas would specify how many students and nonstudents we need to have in the sample and how many people above 40 and how many below 40 we need to have, but it would not specify how many of the students need to be below 40 and how many above, nor would this be done for nonstudents. So, with untied quotas, whether the person is a student and the age of the person would be considered separately. On the other hand, **tied quotas** are quotas where quota variables are combined to create quotas. If we wanted to have tied quotas in the previous example, we would need to define four combined quotas – how many students below 40 we want in the sample, how many students above 40 years of age, how many nonstudents below 40 and how many nonstudents above 40 (Tables 2.7 and 2.8).

Comparing tied and untied quotas, it can be noted that an advantage of tied quotas is that it produces samples that more closely match the properties of the population on the variables used for quotas and thus likely improve the chance that it will resemble the population on all the other variables. On the other hand, tied quotas impose higher demands on data collection personnel during the collection of the sample. When the sampling starts, each entity that is considered for the sample (i.e. study participant in the case of social sciences) will belong to one of the quotas. But as the sampling progresses some quotas will fill quickly, while some will remain unfilled and near the end of sample collection a situation might arise where only entities with certain hard to find combinations traits are required for the sample. When sampling people for participation in a study, this might mean that people recruiting potential participants would need to go through and reject many potential participants because their properties belong to quotas that are already filled, while looking for those rarer combinations that are still needed. The situation becomes harder as the number of variables quotas are based on increases. In extreme cases this

Table 2.7 An example of tied and unted quotas – variable values in the population in the example (fictional data)

Variable values in the population (% of the total population)				
		Student status		*Total*
		Students	*Nonstudents*	
Age	Below 40	14%	36%	50%
	Above 40	1%	49%	50%
Total		15%	85%	100%

Table 2.8 An example of tied and untied quotas – an example of sampling plans (fictional data)

Quotas Sample size = 100 entities (study participants)	
Untied quotas example	*Tied quotas example*
Create a sample of 100 participants (entities) in total consisting of: • 15 students, 85 nonstudents • 50 people below 40 years of age, 50 people above 40 years of age. (Notice how values of the two variables are considered independently i.e. are not tied.)	Create a sample of 100 participants (entities) in total consisting of: • 14 students below 40 • 1 student above 40 • 36 nonstudents below 40 • 49 students above 40 (Notice how the values of the two variables are combined i.e. tied)

can, in practice, create incentives to either falsify the data from the missing categories or finish the sample without them (the former being the more dangerous situation as it usually happens without the knowledge of the researchers running the study and can often not be detected or reconstructed except circumstantially). Due to this, when planning a quota sample, a balance should be struck between the desire to ensure the representativeness of the sample and the need for the sampling procedure to be practically feasible.

Stratified vs. quota sampling. It should be noted that some authors consider stratified and quota sampling to be two variants of the same sampling method with the difference being that after the population is divided into strata i.e. after quotas have been defined (strata being analogues to quotas), if the selection of entities into the sample is done in a probabilistic way, typically through simple random sampling, than that is the case of stratified sampling (some would add -stratified random sampling), while if the selection of entities is done using a nonprobabilistic method, typically through convenience sampling, than that is the case of quota sampling. There are also opinions that stratified sampling is a procedure where we are dealing with subgroups that are administratively or physically or in some other way clearly divided into separate groups, so that entities from different strata are not mixed during the sampling, while quota sampling would include selection on properties of entities that are otherwise mixed in the population. For example, we could choose students of different grades (strata) who we would interview during school hours and we could then select a stratum for an interview simply by choosing which classroom to visit, as all students in the same classroom belong to the same grade (stratum). In contrast, in quota sampling, potential study participants would be individually interviewed to determine which quota they belong to, but people belonging to different groups defined by quotas would be physically together – mixed, in the field. It can be noted that both of these distinctions seem to imply that stratified sampling is a method that requires the population to be limited and organized in some way, because otherwise there would be no list of population members to make random sampling from, nor would it typically be possible to plan for strict separation between members of different strata. But be that as it may, for the purposes of this book, we will take notice of the conceptual overlap between stratified and quota sampling and present both, while recognizing that it might be equally valid to conceptually just present one of these as a variant of the other sampling method.

Purposeful sampling is done when specific entities are selected for inclusion into the sample on purpose. The sample content and the way it is formed is determined in advance in a way that need not follow the distribution of relevant categories in the population. The idea is that certain specific entities of interest be included in the sample such as exceptionally information–rich cases (e.g. for a case study), people with some specific properties relevant for the study, members of groups that were previously found to be extraordinarily predictive of population parameters etc. An example of sampling of this type would be a 2016 study in which a group of researchers purposefully selected two people – a father and a daughter who had the ability to quickly and easily speak backwards for a study in which the researchers examined various biological and psychological factors in an effort to determine what gave them this unique ability (Prekovic et al., 2016). These two persons were selected for the study precisely because they were known to have the unique ability the researchers wanted to study. No other people would do as they do not possess this ability.

Cluster sample is created by dividing the population into groups called clusters and then selecting (typically randomly) clusters the sample will consist of. This can be done in a single stage- when the population is divided into clusters and some clusters, with all their member are selected for study or in multiple stages when the population is divided into clusters, each of which is then divided into subclusters, which can then be themselves divided into subclusters until a level of subclusters is reached with subclusters that are sufficiently small for the purposes of the study. In multistage cluster sampling, after the population has been divided into sufficiently small clusters, the selection of clusters to be included in the sample is made (usually through random sampling) and then all members (or again a certain selection of members) of the selected clusters are included in the sample. A possible example could be if we wanted to obtain a cluster sample of the population of a certain state, we could divide that state into counties, then each of the counties into a number of smaller areas and then select the areas residents of which will be included into the sample. This would constitute a geographical cluster sample. One important requirement of cluster sampling is that clusters need to be homogenous between each other and the content of the clusters needs to be heterogenous. In other words, there should be no clusters that are of themselves too different from the population in general and each should be a small version of the population in a way. This is profoundly different from strata or quotas (in stratified and quota sampling) where groups represented by strata or quotas are selected because of common properties of their members and in this way different between each other. Cluster sampling is often cheaper and easier to do compared to other sampling techniques, yet it still includes procedures aimed at ensuring the sample is taken from different parts of the population and thus reducing the chances of obtaining a sample that differs too much from the population. It can also be used on populations that are not limited or for which we have no membership lists, as long as we can divide them into clusters that cover the whole population.

Snowball sampling is done by selecting the first study participants using convenience or purposeful sampling or some other applicable method and then asking these first participants to recommend further study participants. These other study participants are then asked to recommend further participants and the cycle continues until the needed sample size is reached. Needless to say, snowball sampling is only applicable when entities are people, as it is necessary for the researchers to be able to talk to them and obtain leads on other possible participants from them. Snowball sampling is a method of choice when the goal is to research a specific population whose members are rare in the general human population, but when it can be reasonably expected that members of this specific population are in contact with each other. For example, if we wanted to study the population of bikers, it would not do much good to go from house to house interviewing people and hoping to stumble upon some of these people, as, in most areas, one would need to visit a great many households to find any significant number. On the other hand, as all these people are in contact, through events they all participate in, or clubs or social networks, each of these will likely know a few other people from the same category, who would know at least a few other and so on. Similarly, snowball sampling could be used to collect a sample of for example divers, paintballers, players of various sports that are not very popular, but also of people suffering from certain chronic rare diseases that require specific support types that bring patients into contact with each other or to find survivors of certain event etc.

Systematic sampling is done by creating a list of population members and then sampling every n-th entity on the list. For example, depending on the sample size needed we could sample every 100th or every 50th, 20th etc. This number that is used to define which entity on the list will be selected for the sample is called a "step" of the systematic sample. For example, making a systematic sample with the step 20 means that we select every 20th entity on the list into the sample. When doing systematic sampling it is generally good practice to not start from the first entity on the list, but to first generate a random (pseudorandom) number smaller than the selected step size and then start sampling from it. For example, if we made a systematic sample with a step 30, we would first need to draw a random number between 1 and 29 (smaller than the step). If we, for example, drew number 12, that means that our sample would consist of the 12th, 42nd, 72nd, 102nd etc. entity from the population list. The general rule is that we select entities whose number on the list equals the random number we drew to which we add all possible multiples of the step until we exhaust the list. In general, if the order of entities on the population list is not related to the variables that are studied, systematic sampling can produce results pretty close to random sampling but is somewhat easier. On the other hand, given that creating a systematic sample requires both a list of population members and a way to generate random or pseudorandom numbers cases when systematic sampling is possible are cases when simple random sampling is also possible.

2.9 Issues about samples to be aware of

In most cases, the optimal sample is a sample that is as representative of the population to be studied as possible. However, there are situations in which a representative sample is not optimal. Sometimes the goal of the researchers is to study different parts of the population, some of which are too small to be represented with a sufficient number of entities in the planned sample size. In these cases, researchers may opt for creating a sample in which these parts of the population are **overrepresented** i.e. represented by more entities than would be merited by the size of their part in the population. While such samples are perfectly legitimate, researchers working with such samples should always keep in mind that it is not a proportionate sample and that these groups are overrepresented. This is especially important to keep in mind when making inferences about the whole population as treating a sample with overrepresentation of a certain category of the population as if it were a proportional one could result in seriously misleading inferences.

When it comes to **reporting about the type of sample used in the study** authors should always name their sample according to the sampling technique that best corresponds to the procedure actually used. While sampling procedures can be very different in practice and commonly have certain differences in details from the techniques described in statistics literature, including this book, the procedure should either be described in all detail if it does not correspond to any of the typical procedures or named after the one it corresponds to the most. That said, it can often be found that various researchers name or attempt to name their sample (but get prevented by reviewers before publication!) according to the desired result of their sampling. The most often encountered variant of this is when researchers try to describe their sample as "representative". As described in previous chapters, **there is no sampling technique called "representative sampling"** and also, except in the most trivial cases (very small population, sample the size of the population), **there is no way for the researcher to know whether his/her sampling technique really resulted in a representative sample and to what extent**. This means that whenever authors of research reports describe their sample as "representative", such statement is essentially a misleading one, with authors of such statements claiming something they cannot know. For this reason, we would recommend readers to be vary of any research reports in which the sample is described as "representative", especially if this is not followed by a detailed explanation of the sampling technique actually used. Some authors, especially those doing commercial research or with commercial interests in the study outcomes, might decide to declare their sample representative in the hope of getting the buyer of their research, often a party with only basic statistical knowledge or no statistical knowledge at all to see their research as more reliable or valuable. However, no matter how good or complex the actual sampling procedure used, naming the sample "representative" without actually being able to compare it to the population, is always a misleading practice.

Another common error often found committed by researchers with limited statistical knowledge and those trying to present their research as more valuable is **calling the sample of the study random when it really is not.** It is most often the **case that people who have actually used convenience sampling describe their samples incorrectly as random samples.** Not understanding the actual meaning of random sampling these people will often state that their samples are random because they "interviewed random people on the street" or "interviewed random students in the cafeteria" or "interviewed random patients in the hospital I work in" etc. When reading such examples we need to be aware that there is a difference in the way the word "random" is used in statistics and how it is sometimes used in informal speech. A sample consisting of people one meets on the street is not a random sample, but a convenient one. The same is the case for people in the cafeteria near you or for patients in your hospital. One should always keep in mind the difference between a random and a convenient sample, so when a research study reports that it was done on a random sample, it is a good idea to look up the description of the sampling procedure actually used and verify that it indeed is a random sample and not a case of a convenient sample wrongly named random.

2.10 Levels of measurement

From basic mathematics, we learn of different mathematical operations that can be done with numbers. However, all of us are likely aware of various practical situations where numbers are used, but in such a way that certain mathematical operations make no sense with numbers used in such a way. For example, numbers are often used to mark different apartments in a buildings or different rooms in a hotel. When numbers are used in such a way, we can, for example, say that numbers 1 and number 2 refer to different apartments, meaning that they are not equal, but there would be little point in, for example, adding two room numbers, because apartment number 1 + apartment number 2 do not equal apartment number 3. Such an operation would be pointless. On the other hand, when our numbers are, for example, degrees on the Celsius or Fahrenheit temperature scales, we will all agree that 1°C is colder than 2°C, but we will note that 1°C is not twice colder than 2°C. However, 2°C is 1°C colder than 1°C. In other words, we can subtract one temperature expressed in °C from another and get their difference in °C, but we cannot meaningfully establish ratios between these numbers i.e. divide or multiply them. On the other hand, if the numbers we use are meters or grams, we can easily divide or multiply – a mass of 2 g is twice the mass of 1 g. A 2-meter tree trunk is twice longer than a 1-meter tree trunk. These rules that define which mathematical operations we can and cannot meaningfully do with numbers depending on how they are used are called **levels of measurement.** Probably the most well known and most widely used classification of the levels of measurement was proposed by Stevens (1946), who organized them into four different levels of measurement – nominal, ordinal, interval and ratio. Essentially, these measurement levels differ in the mathematical relations that can be meaningfully established between numbers on each level of measurement. They form a hierarchy with each successive level of measurement allowing for all the relations from the previous level plus some additional ones.

On the **nominal level** of measurement, the only relationship that can be meaningfully established between numbers is that of equality/inequality. Numbers are simply used as either designations of objects or of categories of objects, so, for any two numbers, we can only determine whether they designate the same object or class of objects or a different one. Examples of the nominal level of measurement include when we use numbers to designate different rooms in a hotel or different apartments in a building, when we use numbers to designate players playing in different positions in a football team, but also when we use numbers in a statistical matrix to designate values of variables such as ethnicity (each ethnicity assigned a different number), gender (each gender assigned a different number), various professions a person can work in (each profession or type of profession assigned a different number) or any similar categorical variables with categories between which no other quantitative relationship can be established. Also, we can treat data as nominal even when it is not numbers that are used to designate objects or categories of objects, but words or strings of characters, because the relationship of equality/inequality can equally be established for such designations as the nominal level of measurement does not really require any property that is specific to numbers. Variables with values on the nominal level of measurement are called **nominal variables**.

On the **ordinal level** of measurement, we can also determine whether each value is greater, lower or equal to any other value, in addition to whether the number refers to the same type of object or not. Operators <, > and = can be meaningfully used on this level of measurement. All types of ranks and rankings result in data on the ordinal level of measurement – ranks in a race (we can determine from ranks who had the shorter and who had the longer race time, but not by how much they differ) or in any competition, military rankings (we can determine that captain is a rank above lieutenant, we can also determine that two people with the rank of captain hold the same rank, but we cannot meaningfully tell whether the difference between a lieutenant and a captain is of the same size as the difference between a major and a lieutenant colonel) etc. The ordinal level of measurement is also used in some of the widely known scales such as the Mercalli seismic intensity scale (Wood & Neumann, 1931), various scales for rating tastes and there is also an ongoing debate whether self-rating scales belong to this level of measurement or the next one. Variables with values on the ordinal level of measurement are called **ordinal variables**.

On the **interval level** of measurement, we can also compare sizes of intervals between two values, which means that we can use addition and subtraction (+ and −) in certain ways. Measures on the interval level of measurement have fixed measurement units and, due to this, we can both tell which of the two values is greater and determine the size of the difference between any two values. However, with interval level measures, the start or zero value of the measurement scale is arbitrary, which means that the value of 0 does not indicate the absolute absence of the measured property, but is simply a level that is arbitrarily chosen by the scale creator. For example, we can meaningfully subtract one measure from another to obtain their difference or we can add or subtract a certain number of measurement units to a measure to increase or decrease it. However, we cannot meaningfully add two measures of two objects to obtain the value of an object that has the quantity of the measured property equal to quantities of these two objects combined. An example of the interval level of measurement would be the Celsius and Fahrenheit temperature scales. They both have fixed measurement units and we can meaningfully calculate how much warmer, in degrees, one object is than another. On the other hand, their 0 values are arbitrary – for the Celsius scale 0°C was chosen to be the freezing temperature of water under atmospheric pressure, while 0°F is the temperature at which a brine made of 50% salt and 50% ice melts. If their creators chose some different temperatures, instead of these, to represent 0 degrees of the scale, the scales themselves would function equally well as they do now. Also, if the temperature in a room is, for example 20°C, we can meaningfully increase or decrease it by, for example, 2°C to make the new room temperature 22°C or 18°C. However, we cannot meaningfully say that an object at 22°C contains as much heat as an object on 20°C and an object on 2°C combined. In the same way, we cannot say that an object whose temperature is 4°C is twice as hot as an object whose temperature is 2°C, and the same goes for degrees on the Fahrenheit scale. The interval level of measurement allows the determination of equality of intervals or differences, but not of ratios. In psychology and other social sciences, it is a common practice to consider results of psychological tests (particularly when expressed as standard scores) and results of rating and self-rating scales as being on the interval level of measurement, although there is an ongoing debate whether they all fully conform to the requirements of the interval level or should they be considered ordinal. The main issue in this discussion is whether they fulfill the requirement of fixed measurement units or not. Variables with values on the interval level of measurement are called **interval variables**.

On the **ratio level** of measurement, we can also meaningfully compare ratios, meaning that we can meaningfully multiply or divide the measures (* and /) to determine the equality of ratios. In addition to properties of interval measures, measures on the ratio level of measurement have so-called real zeros i.e. real starting values, meaning that the 0 or starting value of the scale denotes the total absence of the measured property. Due to this we can meaningfully calculate how many times one value is smaller than another. For example, we can meaningfully conclude that an object that is 2 meters long is twice longer than an object that is 1 meter long. An object that weighs 4 g is twice as heavy as an object with a mass of 2 g. These scales are at the ratio level because they also have real zeros – 0 grams denotes a total absence of mass, while 0 meters denotes no distance at all. If you are 0 m from a certain spot, you are standing on that spot and cannot be any closer to it. Examples of the ratio level of measurement include measures for mass in grams, measures of length or distance in meters (or analogous measures of the imperial system!), the Kelvin temperature scale (because 0°K, also known as the absolute zero does represent the total absence of any heat!), but also all results of counting – for example the amount of money on someone's account, the number of doors of a house, the number of livestock a farmer owns etc.

There are various authors who propose additional levels of measurement, apart from the four presented here, such as the absolute level of measurement (it would include results of counting, it is like ratio, but values can only be natural numbers) or log linear (like ratio, but would include measures that are in units that are a ratio of two units, such as m/s or N/m^2). However, as these do not really allow for additional mathematical operations on the measures, in this book, we will stick to Stevens' systematization of the levels of measurement and differentiate only between these four levels. Variables with values on the ratio level of measurement are called **ratio variables**.

One notable **exception to the rules defining the level of measurement are the binary variables or dichotomous variables.** Binary i.e. dichotomous variables are variables that have only two possible values. Because they only have two values **they can be regarded as being on any level of measurement.** It is obvious that the two possible values are different and thus fulfil the conditions for the nominal level. If we refer to these two possible values as A and B (though a binary variable can be a variable with any two values – 0 and 1, true and false, C and D etc.) and we agree that they are different values, there are only two possible ways to rank them – AB and BA. These two rankings actually are the same ordering of values, just in opposite directions and thus binary variables fulfil the requirement for ordinal level of measurement also. The interval level of measurement requires a fixed unit of measurement, which means that it must be possible to divide the variable range into intervals of equal size. With binary variables, there is only one interval and it is always equal to itself thus allowing binary variables to fulfil the conditions of the interval level of measurement. Finally, the ratio level requires that there be a real or absolute zero i.e. that there are no values lower than the zero. Given that on a binary scale there are only two values, we can arbitrarily declare one to be lower (unless one is naturally lower, such as when one of the variables is zero and the other a positive number) then the other and that value will automatically be the lowest possible value, thus allowing binary variables to conditionally fulfil the requirements of the ratio level of measurement. While it is, of course, obvious that binary variables are not really true ordinal, interval nor ratio level measures, these properties allow us to conveniently and in certain appropriate conditions use statistical and mathematical procedures intended for ordinal, interval or ratio measurement level data on binary variables. That said, the results of these procedures on binary data should always be interpreted with taking into account that they were calculated on binary variables and not on true interval or ratio scales. Some statistical procedures, some of which are presented in the later part of this book are even named differently when applied to binary variables in order to reflect that difference.

****

Recognizing the level of measurement the data we are dealing with is on is of key importance when making statistical calculations, because different levels of measurement allow (or disallow) different statistical procedures to be conducted. In general, the higher the level of measurement the greater the number of statistical procedures that can be applied to the data. There are statistical procedures that can be applied to data of all levels of measurement, but there are also statistical procedures that are reserved only for data on a certain level of measurement or above. **For each statistical procedure, there is the minimum level of measurement data needs to be on.** The general rule is that the minimum level of measurement required for a statistical procedure is the one that allows all the mathematical operations needed to calculate that measure to be performed. Due to this, determining the level of measurement of the data is one of the first operations to conduct when examining a set of data and for each statistical procedure that will be described in this book, the measurement level required for its application will also be discussed.

2.11 Continuous and discrete variables

Being aware that variables can have different values and having discussed the issue of the level of measurement, a question arises of how many different values can a variable have. Also, the question arises of how many different values can a variable have in comparison to the number of entities in a sample that is described using that variable. If the number of different variable values is small compared to the sample size, then multiple entities would have the same value on the variable and it could then make sense to describe the sample by counting how many entities have each of the values. If the number of different variable values is large compared to the sample size, then there would be a smaller number of entities with the same value on the variable or it even might be possible that each entity has a different value and, in that case, counting the number of entities with each value would not be very efficient as for many or even every value there would be just one entity having it. Another related question is whether there is a definite list of all the possible variable values or is that list indefinite and also what types of numbers can be used to express these values. This most often comes in the form of the question what can the possible variable values be? When data are on the nominal level of measurement, the answer to this is straightforward – since numbers are just designations for objects or classes of objects, there can be only as many designations as their creator assigns them. In contrast, when we are dealing with interval or ratio data, the question becomes more complicated. Theoretically, between each two points on any interval or ratio measurement scale, we can imagine an infinite number of different values (expressed in decimals or fractions) and this is only limited by the precision of our measurement instruments. In practice, we are, however, free to reduce the number of possible different values in order to simplify the measurement and data recording process by rounding the values or assigning numbers to specific intervals of possible values instead of to individual values.

To address this issue, we introduce the concepts of discrete and continuous variables and corresponding discrete and continual measures. **Discrete variables** are variables than can have only certain, often predefined, values such as only the specific values designated by the researcher or only natural numbers, or only whole numbers, or only tens etc. In contrast to this, values of **continuous variables** can be any real numbers. Discrete variables can be on any level of measurement, while continuous variables obtain their full meaning only on interval and ratio levels of measurement.

This said, the question arises of whether there can really, in practice, be variables whose values can be any real number or can we really, in practice, produce continuous variables. The answer to this is obviously no. Continuous variables are a theoretical concept, but in practice we are always limited by the precision of the measurement instrument we use to obtain measures. As there are no measurement instruments with absolute, infinite precision, we cannot really produce any real number as a result, but are limited to the minimum differences between variable values that the instrument we are using can differentiate between. So, in practice, there are no continuous variables, just discrete variables with broader and narrower categories or, in other words, with a larger and smaller number of categories or possible discrete values. However, when the possible number of categories becomes sufficiently large and we are dealing with an interval or a ratio variable, we may consider and do consider such measurements to approximate continuous measurements sufficiently well, to allow such measurements to be practically treated as continuous.

2.12 **Let us apply what we learned so far!**

Let us now try to apply what we have covered in this chapter through a couple of exercises. Please refer to the start of the book for the general instruction for completing the exercises. Our suggestion is that you first read the excerpt and the statements and provide your own answer. You may write it in the Answer column, and after that look up the answers and compare your own answers with them.

Exercise A. Variables, levels of measurement, sampling

John is doing a study in which he plans to explore whether personality traits of people working as football referees differ from the general population. He will do that by comparing the results of a group of football referees on the HEXACO PI-R personality test to norms of this test for the general population. He plans to acquire a sample of football referees by taking the list of all football referees in the state and than using a random number generator to select those that will be included in his sample. He plans to obtain a sample of 600 football referees is this way. In this sampling procedure, a referee that is once included in the sample will not be considered in the drawing of the following participants (i.e. one referee can be included in sample only once).

The testing of these referees itself will be done using the HEXACO PI-R personality inventory that produces results on 6 different personality dimensions. These test scores are considered to be on a scale that has a fixed measurement unit, but an arbitrary 0 i.e. test-takers cannot have a value of zero and the minimum possible value is equal to the number of items in a specific scale. John will also record the profession of the referees outside sports (most of them have another profession outside of being a referee), their gender (with the options being male, female and other) and the number of football matches they have refereed during their career.

Ellen is doing a study at the same time as John. Ellen plans to do her study herself and she will only study football referees living in the state who refereed at least one international match in the last year. She plans to collect her sample by finding a couple of referees at the football association and than asking them to recommend their colleagues – other referees. She will than test these recommended referees and ask them to recommend other colleagues and she will continue like that until she collects the needed sample size. All referees that Ellen asked to participate agreed to participate.

Please assume that in all the described situations obtained distributions are identical to theoretical distributions that are expected in situations of the given type.

A	Statement:	Answer
A1.	If all referees invited by John agreed to participate in John's study, John would have a convenient sample.	
A2.	Ellen plans to conduct her research on a snowball sample.	

A	Statement:	Answer
A3.	All 600 football referees selected by John have agreed to provide all the data he required.	
A4.	John's sample will be representative of the population.	
A5.	HEXACO PI-R scores are on the ratio level of measurement.	
A6.	If all referees in Ellen's study turned out to be male, than gender would be a constant in her sample.	
A7.	John is conducting sampling without replacement.	
A8.	The number of football matches a referee has refereed during his/her career is a discrete variable.	
A9.	Ellen's sampling procedure is guaranteed to result in a representative sample.	
A10.	In John's study, gender is a binary variable.	

Exercise B. Variables, levels of measurement, sampling

Katie is conducting a study that has the goal of finding out how satisfied the county residents are about a new application for the issuance of legal documents her county is using. This application is intended to serve all residents of that county and, therefore, all of them are the target group for this study. It is an urban county with roughly 20 thousand residents. She conducted the study by going into a park near her house and asking the passers-by there to fill-in her questionnaire. She continued with this until she collected responses from 200 people, which is the number of people she planned to have in her sample.

Her colleague, Isabella conducted the same study, but she created her sample by taking the official database that contains the list of all residents of the county and then asked every 100th resident on the list to participate in the sample.

Both Katie and Isabella asked the residents to rank the various services that the county provided and then noted the rank that their application received in comparison to other services offered by the county. They also recorded the age of the study participants. Finally, they divided the county into several areas, assigned a number to each, and recorded the area in the county each respondent lives in.

B	Statement:	Answer
B1.	Katie's sample is representative of the population.	
B2.	Katie collected a random sample.	
B3.	Isabella applied systematic sampling to create her sample.	
B4.	Isabella's sample is more representative than Katie's.	
B5.	Age of study participants is a variable on the ratio level of measurement.	
B6.	The variable through which the quality of the application is compared to other county services is on the interval level of measurement.	
B7.	The variable indicating the area a study participant lives in is on the nominal level of measurement.	
B8.	Isabella also had approximately 200 study participants if everyone she selects decides to participate.	
B9.	There were more males than females in Isabella's sample.	
B10.	Age of study participants is a primordial variable.	

Let us now consider the answers:

A1 − false. The story states that John plans to collect his sample by using a random number generator to select sample participants from the population list. That is a description of a simple random sample, not a convenient one.

A2 − true. Ellen will be asking her initial participants to recommend further study participants and this is a snowball sample.

A3 − unknown. The statement refers to what happened during the data collection and this is not covered by the story.

A4 − unknown. While a simple random sample John is collecting is a good sampling procedure, we can never know whether the sample is representative unless we are able to compare it to the population. However, if we had population data, we would not need a sample in the first place.

A5 − false. It is clearly stated in the story that these scales do not have a real 0 value, hence cannot be on the ratio level of measurement.

A6 − true. Yes, if all entities in a sample have the same value of a certain property, that property is a constant, not a variable. For a property to be a variable in a sample, entities have to have different values on it.

A7 − true. It is stated that a referee once included in the sample is not considered for inclusion again. That is the definition of sampling without replacement.

A8 − true. While different referees will have different numbers of refereed matches, these values can only be natural numbers, making this a discrete variable.

A9 − false. There is no sampling procedure that is guaranteed to result in a representative sample.

A10 − false. It is stated in the story that gender will have three categories. A binary variable is a variable with two categories.

B1 − unknown. Nothing in the story indicates whether it is representative or not and no sampling technique can guarantee that the sample collected will be representative. Also, even if the sampling procedure is convenient, it does not mean that it cannot result in a representative sample.

B2 − false. Katie's sample is convenient. Although we could likely, in everyday talk, say that she interviewed "random people on the street", such approach to sampling collection is called convenient sampling and has nothing to do with random sampling.

B3 − true. She applied systematic sampling from the list of residents with the step of 100.

B4 − unknown. There is nothing in the text that would allow us to make any definite conclusion about which sample is representative. Again, sampling technique applied is no guarantee that the resulting sample will be more representative than any other sample.

B5 − true. Yes, age is a ratio variable. A 20-year-old person is twice the age of a 10-year-old person. That is ratio level.

B6 − false. The variable in question is a ranking. Rankings are the ordinal level of measurement.

B7 − true. Areas are designated by numbers and the only thing we can meaningfully say about any two persons in regard to this variable is whether they live in the same area or in different areas. Numbers are designating categories. That is the nominal level of measurement.

B8 − true. If the county has 20 000 residents and she uses systematic sampling with step 100, that is 20 000/100 = 200. Give or take a few respondents in the case that she did not start from participant number 1, or that there might be some residents above or below 20 000, as the text says that there are roughly 20 000 of them.

B9 – unknown. There is no reference to male to female ratios in the samples in the text.

B10 – meaningless. There is no such thing as a primordial variable.

Notes

1 It should be noted that although there exist cases in the world where two heads live attached to the same body, such cases are legally considered as two persons, Siamese twins, sharing the same body, not a single person with two heads. Therefore, the number of heads per person is a constant even if we take such situations into account.
2 It should be noted that there are situations when a study is focused on learning more about specific parts of a population and in those situations an ideal sample might be a sample that differs from the general population, usually in the number of entities representing certain parts of the population, for example, with some being overrepresented, but this again comes down to the need for the sample to be representative of the larger set we wish to infer about based on the sample. We should just remain aware of which population did we create our sample to be representative of.

References

Akhshani, A., Akhavan, A., Mobaraki, A., Lim, S. C., & Hassan, Z. (2014). Pseudo Random Number Generator Based On Quantum Chaotic Map. *Communications in Nonlinear Science and Numerical Simulation, 19*(1), 101–111. 10.1016/j.cnsns.2013.06.017

Desai, V. V., Deshmukh, V. B., & Rao, D. H. (2011). Pseudo Random Number Generator Using Elman Neural Network. *2011 IEEE Recent Advances in Intelligent Computational Systems*, 251–254. 10.1109/RAICS.2011.6069312

Efron, B. (1979). Bootstrap Methods: Another Look at the Jackknife. *The Annals of Statistics, 7*(1), 1–26. 10.1214/aos/1176344552

Good, P. I. (2006). *Resampling Methods A Practical Guide to Data Analysis Third Edition.* Birkhauser. www.birkhauser.com

Liu, J., Liang, Z., Luo, Y., Cao, L., Zhang, S., Wang, Y., & Yang, S. (2021). A Hardware Pseudorandom Number Generator Using Stochastic Computing And Logistic Map. *Micromachines, 12*(1), 1–12. 10.3390/mi12010031

Miller, R. (1974). The Jackknife - A Review. *Biometrika, 61*(1), 1–15.

Prekovic, S., Filipović Đurđević, D., Csifcsák, G., Šveljo, O., Stojković, O., Janković, M., Koprivšek, K., Covill, L. E., Lučić, M., Van Den Broeck, T., Helsen, C., Ceroni, F., Claessens, F., & Newbury, D. F. (2016). Multidisciplinary Investigation Links Backward-speech Trait And Working Memory Through Genetic Mutation. *Scientific Reports, 6*, 1–15. 10.1038/srep20369

Stevens, S. S. (1946). On the Theory of Scales of Measurement. *Science, 103*(2684), 677–680.

Tošić Radev, M., & Hedrih, V. (2017). Psychometric Properties Of The Multidimensional Jealousy Scale (Mjs) on a Serbian Sample *. *Psihologija, OnlineFirst*, 1–14. 10.2298/PSI170121012T

Wood, H., & Neumann, F. (1931). Modified Mercalli Intensity Scale of 1931. *Bulletin of the Seismological Society of America, 27*(4), 277–283. https://scits.stanford.edu/sites/g/files/sbiybj13751/f/277.full_.pdf

3 Descriptive statistics

Processing statistical data on a sample typically starts with first describing the sample and then making inferences about the population based on those descriptions. Statistical procedures and indicators used for describing samples constitute the area of statistics called descriptive statistics. The most basic of such procedures and indicators will be presented in this chapter.

3.1 Distribution

A list or a description of values of entities in a sample is called a distribution. A distribution of a variable can consist of a list of all values of all entities in the sample or can consist of a depiction of all possible values with information about how often entities with each specific value are encountered in a sample. When variables are discrete or, practically, when the number of possible different variable values is small, a distribution can be described by listing how many entities have each of the values. Such distributions are called **discrete distributions.** If the number of possible different values is too large to be presented in the former way, than values can be grouped into broader categories and the distribution can be presented by describing how many entities there are in each category. If we are describing the distribution of values of a continuous variable on a sample, we are dealing with a **continuous distribution.** A continuous distribution can be represented using a **probability density function,** which is a function whose value at any point can be interpreted as a relative likelihood that a randomly taken entity from the sample will have a value close to that. A description of a continuous distribution using a probability density function is usually created by dividing the span of possible values of the continuous variable into a number of intervals and then examining the number of entities found in each interval. These relative numbers are than taken to represent the likelihood of a randomly selected entity falling into a certain interval, which is then used to construct the function.

For making a basic description of a distribution, we need to introduce the following concepts:

- **Frequency** – is the number of entities that have a certain value of a variable. We obtain the frequency of a certain variable value or of a certain variable category by counting the total number of entities in the sample that have that particular value i.e. the number of entities that belong to the particular category whose frequency we are calculating.

DOI: 10.4324/9781003107712-3

- **Proportion** – is the share of entities with a certain value i.e. that belong to a certain category in the total number of observed entities. Proportion is calculated by dividing the frequency of the category whose proportion we wish to calculate with the total number of entities in the sample (proportion = frequency/total number of entities in the sample). Proportions are on a scale from 0 to 1. If the proportion of a value is 0, that means that there are no entities with that value. If the proportion is 1, that means that all entities in the sample have that particular value, i.e. that the examined variable is a constant (and not a variable, because a variable requires that entities have different values, when all entities have the same value, we are dealing with a constant).
- **Percentage (%)** – is a proportion multiplied by 100. It is the same as proportion just rescaled to a span between 0 and 100. Percentage is denoted with the sign %. Percentage is essentially the same type of statistic as the proportion indicating the relative share of a certain category or value in the total sample, only using a number range that might be more convenient for practical use and communication (small, often whole number, instead of decimals).

3.2 Percentiles and other quantiles

Sometimes there is a need to mark specific positions on a distribution of a non-nominal variable. This is most commonly done through the use of percentiles and percentile ranks.

- **Percentile** represents a point on the distribution below or in line with which lie the values of a certain percentage of entities from the sample. In other words, it is a value of the variable that is higher or equal to a specific percentage of values of entities in the sample. A percentile is named based on the percentage of entities that have values lower than or equal to the value of that percentile. For example, 50th percentile is a value that is higher than or equal to values of exactly 50% of the sample on the given variable. 100th percentile is the highest value in the sample. 0 percentile is the lowest value in the sample. 20th percentile is the value below or in line with which lie the values of exactly 20% of entities in the sample (while 80% have values higher than that percentile). Percentiles are usually calculated by ordering the entities in the sample in an ascending or a descending order according to their values on the considered variable and then going from the smallest up finding the value for which no more than the required percent of entities has lower values and at least the required percentage of data has smaller or equal values to it. This is called the **nearest-rank method** of determining percentiles.
- **Percentile rank** is the percentage of entities in a sample that have lower or equal values to the value of the considered entity. Therefore, percentile rank is the property of a specific entity. It is closely related to percentiles, however, percentile is a value of the variable, a point on the distribution, while percentile rank is a property of an individual entity. In practice, we would say that, for example, 30th percentile is a certain value (a certain number). On the other hand, for an entity whose value on the considered variable equals 30th percentile, we would say that it holds the rank of 30. For example: "30th percentile of our sample on the variable X is 42 (variable value)", but "John scored 42 on the test, thus his percentile rank is 30".

Percentiles and percentile ranks are the most commonly used markers of position in a distribution (i.e. how big certain value is in comparison to values of entities in a sample) and these are based on percentages i.e. division of the sample into 1/100th part fractions. However, there is no universal rule stating that position in the distribution must be indicated using 1/100th part fractions. This means that any other fraction i.e. dividing the whole distribution into any other number of equal parts, is equally valid. The general name for such divisions of the distribution into a certain number of equal parts is **quantiles** (Harding et al., 2014), also often called **n-tiles** where n stands for the designation of the number of parts the whole distribution is divided into. Percentiles are division of the distribution into 100 equal parts, but it can be any number of parts with equal validity. Other commonly used quantiles aside from percentiles are:

- **Quartiles** – that divide the distribution into 4 equal parts. There are therefore 4 quartiles. the 1st quartile corresponds to the 25th percentile, the 2nd quartile to the 50th percentile, 3rd equals the 75th percentile and the 4th quartile equals the 100th percentile. It should be noted that some authors use quartiles not as indications of points in the distribution, but to denote intervals on the distribution. In such a system the 1st quartile refers to the quarter of the sample with the lowest scores, 2nd quartile refers to variable values that are between the 25th and the 50th percentile, the 3rd quartile refers to values that are between the 50th and the 75th percentile and the 4th quartile would be comprised of the top quarter of the sample with regard to the values on the considered variable.
- **Quintiles** – divide the distribution into 5 equal parts. All the details described for the quartiles hold equally with quintiles with the only difference being that the sample is now divided into 5 equal groups instead of 4.
- **Deciles** – divide the sample into 10 equal groups. The 1st decile corresponds to the 10th percentile, the 2nd to the 20th percentile and so on. Deciles are also sometimes used to denote intervals and in this case the 1st decile refers to the 10% of entities in the sample with the lowest values, 2nd decile refers to entities whose values are between the 10th and the 20th percentile and so on.

Percentiles, but also other quantiles are used whenever there is a need to denote the position of an individual on a distribution of a variable that is at least on the ordinal level of measurement. For example, percentiles are used in psychology as the primary statistic for interpreting the level of expression of various psychological traits measured using psychological tests in the scope of so called norm-referenced approach to interpreting test results (for more details see e.g. Hedrih (2020)). In education, quantiles can be used to divide a group of students based on their test achievement into several groups, in clinical assessment percentiles are often used to differentiate between diagnostically relevant results and results that are considered "normal" or typical. In the economy, quantiles are often used to divide companies working in an area into categories according to their business performance. In many other areas, division of the sample into multiple groups based on values of a variable of interest or denoting specific positions is useful/ needed and these are the needs that quantiles solve.

The **use of quantiles requires that the data be at least on the ordinal level of measurement**.

*****

While frequencies and percentages and measures derived from them may be adequate for describing distributions of nominal variables and distributions of variables with a small number of categories, for higher levels of measurement it is more typical to describe a sample by calculating a statistic that denotes the most common or typical value (or values) of the variable and another statistic that describes the magnitude of differences between values of entities in the sample. The former types of statistics are called measures of central tendency and the latter are measures of variability or measures of dispersion.

3.3 Measures of central tendency

Measures of central tendency are a category of statistics whose purpose is to indicate the central tendency of entities in the sample with regard to values on the considered variable. Ideally, they represent a point in the distribution/the value of the variable that entities are the most likely to have or around which their values tend to cluster (on higher levels of measurement). The most widely known measures of central tendency are arithmetic mean, also known as the average, median and mode.

Mean, arithmetic mean or the average is calculated by adding up the values of all entities in the sample and dividing it with the number of entities in the sample:

$$Mean = \frac{\sum_{i=1}^{n} X_i}{N}$$

In scientific literature, it is typically denoted with the letter M or the Greek letter μ. The arithmetic mean can be meaningfully calculated for data that are on the **interval or ratio levels of measurement**. However, the formula for calculating the arithmetic mean can be used in some cases on ordinal data, but the statistic obtained in that way is called the mean rank. **Mean rank** can be a useful statistic when we have multiple groups that are ranked together on a joint rankings list and in that case mean rank would indicate whether members of that particular group tend to be in the upper or in the lower part of the list, thus showing whether their values tend to be among the higher or among the lower values in comparison to the rest of the sample. In most cases, there is little point in calculating mean rank for the whole sample, because if our ordinal data is a ranking list, the mean rank will simply be a half of the number of entities in the sample (if entities are ranked from 1 to the last) or a number that does not mean anything in particular if the system of ordinal values works in some other way due to the absence of the fixed unit of measurement.

One very important feature of the mean as a measure of central tendency is that it is affected by values of all entities in the sample, taking, in a way, all entities into account. However, this same feature has an important negative side which is the fact that the value of the mean can be greatly affected by a single extreme value or just a small cluster of them. In practice, this means that, for example, if we wanted to calculate the average income of a group of people and this group consisted of 100 people earning 100$ per year and one person earning 1,000,000$ per year, the mean income of the group would be 10,000$ per year. Now, the mean of 10,000$ obviously is useless and quite a misleading statistic for this group as it is a value that nobody in the group has, thus missing the primary purpose of a measure of central tendency, which is to indicate a typical or central tendency of the group. There is also the running joke popular among students and teachers of statistics that says that if we have two groups of people and one is eating cabbage, while the other is eating steaks, this type of calculation would tell us that these two groups are eating cabbage casserole with meat on average, which is obviously wrong. There are also known examples of various governments throughout the world (mis)using the mean to try to present salaries in the country as higher than they actually are by presenting the mean salary as an official statistic, while downplaying or neglecting the fact that the average salary is greatly moved upwards by a small number of extremely high salaries (just as in the example above). In such situations, the average salary is a number that is higher (often much higher) than what most of the people actually earn. This is the reason why an **additional requirement for calculating the mean** is that the values of entities be distributed in such a way that **the values of most are grouped around a central focal point** and that there are fewer and fewer entities as values go further away from that central point. In a later chapter of this book, theoretical distributions will be discussed and we will see that many authors consider a so-called normal distribution of the data to be a prerequisite for a meaningful use of the mean.

Apart from the arithmetic mean, similar measures of central tendency include the **harmonic mean** and the **geometric mean**, but we will rarely see these statistics used in social and behavioral sciences. The **harmonic mean is calculated by dividing the number of entities in the sample with the sum of reciprocals of values of entities in the sample:**

$$Harmonic \ mean = \frac{N}{\sum_{i=1}^{n} \frac{1}{X_i}}$$

Geometric mean is **the Nthroot of the product of values of all entities in the sample**, where N is the number of entities in the sample. It is good to be aware of the existence of these measures of central tendency, but we will rarely, if ever, encounter either of these measures applied in practice in social and behavioral sciences (they have their applications in other sciences, though).

Median is the center of the distribution, a point above and below which lie the values of 50% of the sample in each direction (50% above the median and 50% below the median). In other words, **median is the 50thpercentile** or the 2nd quartile. The meaningful calculation of the median requires that the data be **at least on the interval level of measurement**, although there are situations when a median can be meaningfully used on **the ordinal level of measurement** (calculating medians of subgroups for comparison).

In some ways, the features of the median are the opposite of those of the mean – median is also influenced by all entities in the sample, only not by their values, but only by their position in regard to the central part of the distribution. In this way, the median is not at all affected by cases with extreme values, no matter how extreme. On the one hand, this is a good thing, as the value of the median will always be a value that some entity in the distribution does have. In the previous example, with 100 people earning 100$ per year and one person earning 1,000,000$ per year, the median would be 100$, because that is the middle of the distribution i.e. the 50th percentile. Now, 100$ is clearly a better assessment of the earnings of this sample, because this measure of central tendency clearly indicates the income level that most people in the sample have. However, if we realistically considered a group such as this, in which, in a group of people with practically zero earnings, there is one earning a million, we would need to conclude that neglecting the fact that such a person exists would also be assessing the group incorrectly – a group of 101 people of which everyone earns 100$ will likely be quite different from a group in which there is a person with a 1,000,000$, precisely because of the existence of that one person. Also, if our sample consisted of 51 people earning 100$/year and 49 people earning 1,000,000$ per year, the median would still be 100$ and we can easily agree that this would be an even worse assessment than the one in the previous example. Due to this, the most meaningful use of the median is also in the situations when most of the sample is grouped around one central location with the number of entities becoming smaller and smaller as we move away from the group of central values. Officially, the requirements for calculating a median are less strict than those for calculating the mean (as we will see in later chapters), however, the purpose and usefulness of the median as an indicator of central tendency is quickly defeated as the shape of the distribution moves away from what is previously described.

Mode is **the most common value in a sample** i.e., the value with the highest frequency. A sample can have more than one mode. This happens in situations when two or more of the most frequent values also have the same frequency. Also, when reporting a mode and there are two or more values that are strikingly frequent in comparison to the frequency of other values, many authors will report both or all of such values even when their frequencies are not exactly the same i.e. when technically, just one of them is the mode, but the authors reason that omitting to report on the other most frequent values would not represent the properties of the sample on the considered variable adequately. Mode can be calculated for data on the **nominal level of measurement** and it is actually the only commonly used measure of central tendency that can be meaningfully calculated for nominal data. Mode can also be calculated on higher levels of measurement, but such a calculation is meaningful only if we are dealing with discrete measures with a small number of different values compared to sample size or if the measures have been merged into a relatively small number of groups based on value intervals (by specifying the ranges of values that will be designated as the same category, and keeping the number of such categories small). If the number of different values of the variable the entities have is large compared to the sample size, than we will have a situation where most of the values will have a frequency of 1 or will be in a low digit value and that is the situation where the category with the highest frequency will likely be due to random chance and not indicative of any central tendency of a sample. An example of this situation would be, for example, a group of people whose height we would be measuring with a very precise instrument and reporting it in microns (1 micron is 1/1000 of a millimeter). With such precision and a relatively small sample, it is quite likely that we would not even find two persons with exactly the same height to the last micron and even if we did find two such people and declared it a mode, such a mode would indicate little about the general tendencies in the sample. However, if we rounded the height values into decimeter categories – for example, making those below 1,5 meters the first category, those between 1,5 meters and 1,6 meters the second, those between 1,7 meters and 1,8 meters the third category and so on, a mode would be a much more meaningful indicator of the central tendency of the sample regarding the height.

The mode is sometimes also called the **modal value**. We can see it used in literature in formulations like "Mode of the sample is 23", but also "the modal value is 23".

3.4 Measures of variability

Measures of variability show the level of variability of entities in a sample i.e. the magnitude of differences between entities in the sample. The most commonly used measures of variability are standard deviation/variance, quartile deviation, range and the coefficient of variation.

Standard deviation is calculated by subtracting the sample mean from the value of each entity, squaring those differences, then calculating the average of these squared differences and then taking the square root of the result. In scientific literature, **standard deviation is typically denoted with letters SD or the Greek symbol σ.** Standard deviation can be **meaningfully calculated only if the data is on the interval or ratio levels of measurement.** It is typically paired with the mean when describing the sample – when we use the mean to describe the central tendency of a sample, we use the standard deviation to describe its variability.

The formula is:

$$SD = \sqrt{\frac{\Sigma\,(X - M)^2}{N}}$$

or

$$SD = \sqrt{\frac{\Sigma\,(X - M)^2}{N - 1}}$$

where M is the sample mean, X is the value of an individual entity, Σ is the sum operator, meaning that after we have subtracted the value of each individual entity from the sample mean and squared the result, we calculate a sum of all those squared differences. N refers to the number of entities in the sample. Often, N in the formula is replaced with N-1 which represents a value called the number of **degrees of freedom**. In statistics, degrees of freedom indicate the number of values that are free to vary without violating given constraints. It is calculated by subtracting the number of relations from the number of observations (number of entities in the sample in this case). For calculating the standard deviation, it is always N-1. That said, for most practical purposes found in social sciences and most other sciences and given the typical sample sizes, **it most often makes no practical difference whether we use the number of entities or the number of degrees of freedom in this formula** i.e. whether we use the first or the second formula for calculating the standard deviation. For most practical purposes, dividing something with 200 does not produce a result that is much different than if it were divided by 199 (for a sample of 200 entities) or even better, there will be very little difference in the result if we divide with 1300 compared to dividing with 1299 (for a sample of 1300 entities).

Just like the situation with the mean, an important feature of the standard deviation is that its size is affected by the magnitude of deviation from the sample mean of all entities in the sample, and deviations increase the value of the standard deviation regardless of the direction they are in (whether the individual value is greater or lower than the mean), because the sign of this difference is lost when the result is squared, so all the squared differences are positive regardless of the sign of the difference itself. Like the mean, the standard deviation can also greatly be influenced by entities with extreme values – individual extreme values can visibly increase the size of the standard deviation, so all the pros and cons already discussed about the use of mean apply here as well. The requirement about the shape of the distribution needed to make the calculation of the mean meaningful, applies equally to the calculation of the standard deviation. It should also be noted, that when the considered property is a constant, the standard deviation will be zero.

Variance - calculating the standard deviation includes taking the square root from the average of squared differences. If we omit that last step and do not take the square root, but just calculate the average of squared differences between values and the mean, we obtain a statistic called **variance.** In other words, **variance is the square of the standard deviation and vice versa** – standard deviation is the square root of variance. That is why, variance is often denoted like just the square of the standard deviation – σ^2, but it is as often denoted with just the letter V or letters VAR. As a measure of variability, variance functions the same as the standard deviation. Given that it is just the square of the standard deviation, standard deviation and variance are always in perfect concordance relative to conclusions about the variability that can be drawn from them. However, variance is a statistic greatly used in various multivariate statistical techniques (not covered in this book) and various advanced statistical procedures.

Median absolute deviation is the **median of the absolute values of deviations of entities in the sample from the sample median**. In other words, it is the median of differences between individual values of entities and the sample median when we disregard the direction of the difference from the median (i.e. whether the value of a particular entity is higher or lower than the median). It is calculated by subtracting the value of the median from each individual value, removing the sign from these results (+ or – or in other words turning them into absolute values) and then calculating their median, which is then called median absolute deviation. The relationship between the median and the median absolute deviation is similar to the relationship between the mean and the standard deviation. However, at the moment this book is written, it is relatively rare to encounter the median absolute deviation used in presenting results of scientific research, particularly in the social and behavioral sciences. The median absolute deviation can be meaningfully calculated when data is at least on the interval level of measurement.

Quartile deviation is **one half of the difference between the 75th and the 25th percentile** or, in other words, between the 3rd and the 1st quartile (hence the name – quartile deviation). The difference between the 3rd and the 1st quartile is called the interquartile range, so quartile deviation can also be defined as one half of the interquartile range. The quartile deviation **roughly shows the magnitude of dispersion of the middle part of the distribution**. It is typically used in conjunction with the median. Quartile deviation is fully meaningful when the data is at least on the interval level of measurement, as a fixed unit of measurement is needed to make the meaning of the dispersion magnitude fully meaningful. However, there are situations when it can also be meaningfully used on the ordinal level of measurement.

Like the median, quartile deviation is influenced by the value of each entity relative to other entities in the sample, but not by what exactly that value is. Due to this, unlike the standard deviation, quartile deviation is not influenced by extreme values of entities on the considered variable.

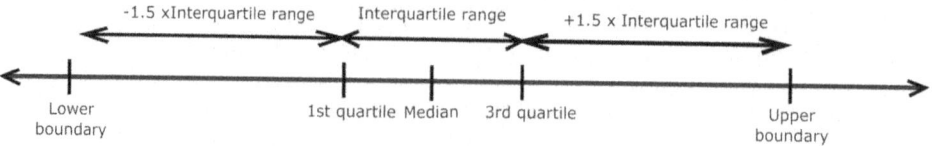

Figure 3.1 An illustration of the 1.5 interquartile range rule on a number line. 1.5 times the span of the interquartile range is added to the 75th percentile (i.e. 3rd quartile) and the same value is subtracted from the 25th percentile (i.e. 1st quartile). Entities outside the range defined by these two values are considered outliers.

Range is simply the difference between the highest and the lowest value in the sample. It shows the interval inside which lie the values of all entities in the sample. While knowing this interval is certainly useful, the problem with the range is that it is solely influenced by the two most extreme cases in the sample – the most extreme low value and the most extreme high value. The problem with this in practical research is that these extremes can often be the result of some erroneous or invalid measurements or also sometimes, when the researchers are not sufficiently careful, of an error in data entry. If such a situation occurs, the range becomes meaningless. Due to this, it is sometimes useful to calculate a so-called **trimmed range** i.e. range calculated on a sample from which a couple of entities with the highest and the same number of entities with the lowest values have been excluded. Sometimes, trimmed range is also calculated by excluding from the sample the entities whose values differ substantially from values of other entities in the sample i.e. whose values differ by more than a predefined value from the central body of the distribution or from a certain point on the distribution. Such entities are called **outliers.** For example, one way to decide which entities are outliers is the so-called 1,5 interquartile range rule or the 1,5 IQR rule. It essentially states that we should multiply the interquartile range by 1.5 and then add that value to the 3rd quartile and subtract it from the 1st quartile. All entities whose values fall outside the interval defined by these two values are considered outliers according to this rule. (Figure 3.1).

3.5 How can a distribution be represented?

Other than simply making a list of all entities in the sample with their values, i.e. as a vector, distribution is typically represented in literature either as a table in which all different values of the variables are presented with the frequency, proportion or percentage of each value or through a number of graphical means.

Discrete distributions can be represented through pie charts, line diagrams and similar graphical depictions, while continuous distributions are typically presented using histograms, boxplots and graphical presentation of the probability density function (Figures 3.2–3.8).

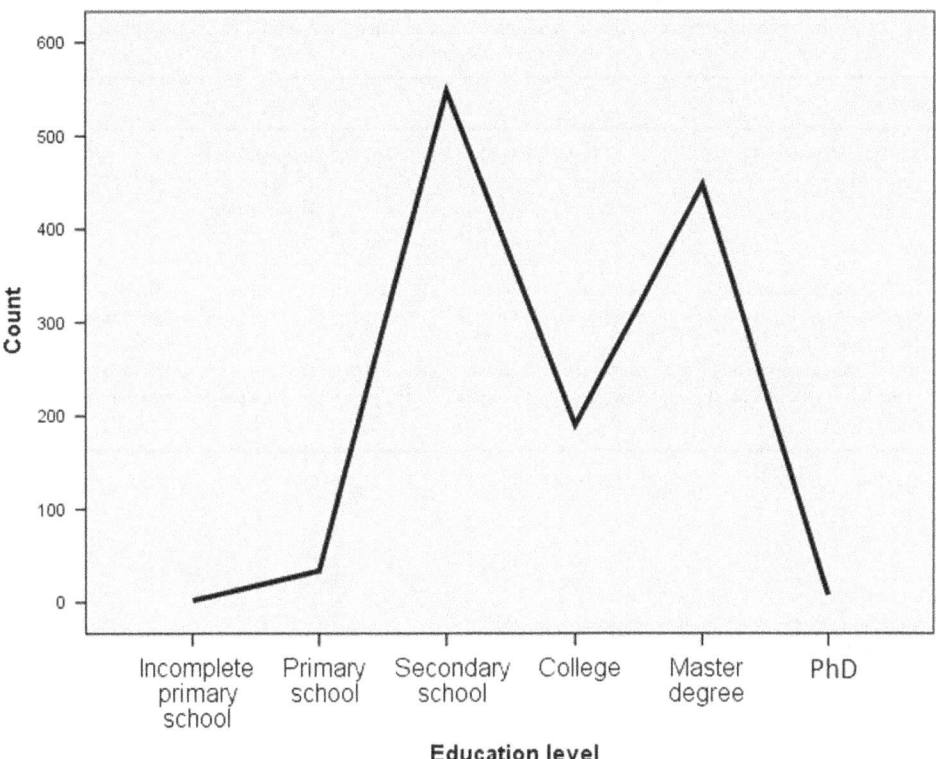

Figure 3.2 A representation of a distribution through a line graph – distribution of education levels of the sample from a study of work-family relations in Serbia. The vertical axis represents frequencies i.e. the number of study participants in the category, while different education categories are listed on the horizontal axis (Hedrih, 2017). We can see from the chart that the most frequent education level in the sample is secondary school, with more than 500 study participants reporting this education level, making secondary school the mode of the variable Education level.

Table 3.1 A representation of a distribution through a table with data on percentages of entities in each category. The data presented in the table are answers of participants in a study by Hedrih & Hedrih (2012) on potential sperm donors (donors for the purposes of artificial insemination) who were asked about the properties of people whom they would be willing to give a sperm donation to. The first column are variables i.e. questions they were asked, while the subsequent columns in that row are values of these variables – possible answers offered and percentages of study participants who gave each of the answers. For example, we can see from the table that 47.8% of participants stated that acquaintance (with donation recipients) is not important for their decision to make a sperm donation. Also, from the third row, we can see that 69.9% of study participants stated that they would give consent for their sperm donation to be used by an (unmarried) heterosexual pair, while 90.3% stated that they would give consent that a married couple be the recipient of their sperm donation

Question	%					
Who would you make a donation to?	people I am acquainted with	people I am not acquainted with	both	acquaintance not important		
	5,5	26,5	20,2	47,8		
Which categories would you make a donation to? (% checked)	married couple	heterosexual pair	lesbian pair	widow	single woman	divorced woman
	72,3	32,0	12,9	27,7	40,3	29,0
Would you give consent for your sperm being used by... (% yes)	married couple	heterosexual pair	lesbian pair	single woman	divorced woman	
	90,3	69,9	22,2	61,2	54,1	

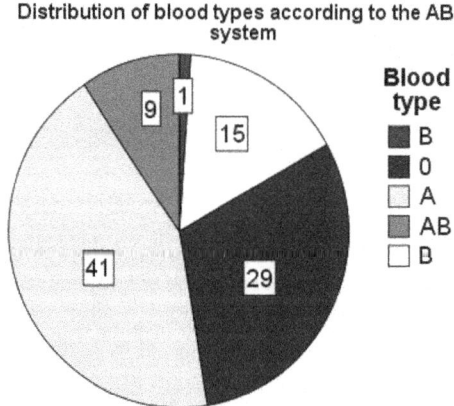

Figure 3.3 A representation of a distribution through a pie chart – distribution of blood types according to the ABO system, a nominal variable. Pie slices represent the share of the study participants with a specific blood type in the sample. Numbers on the slices are frequencies i.e. the number of study participants in that category. The pie chart is based on data from the study by Hedrih et al. (2018) on the influence of hand microtrauma on pathogenesis and progression of hand and neck osteoarthrosis. We can see from this particular chart that most study participants – 41, had the A blood type, making A blood type the mode of this sample on this variable. The least common was the B blood type, with only 1 participant having that type.

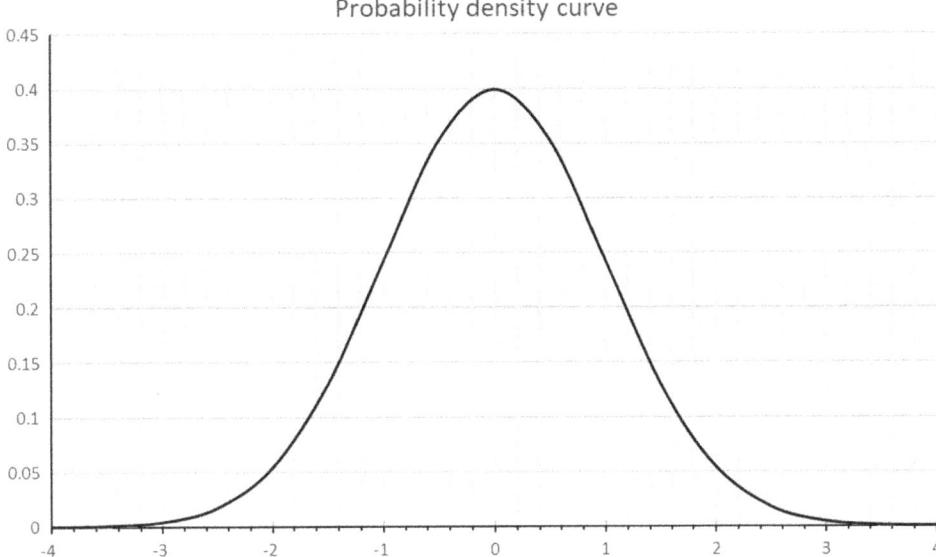

Figure 3.4 A representation of a distribution of an interval level of measurement variable through a probability density function – the horizontal axis represents variable values, while the vertical axis represents probabilities (given as proportions) of a randomly selected entity from the population have that particular value.

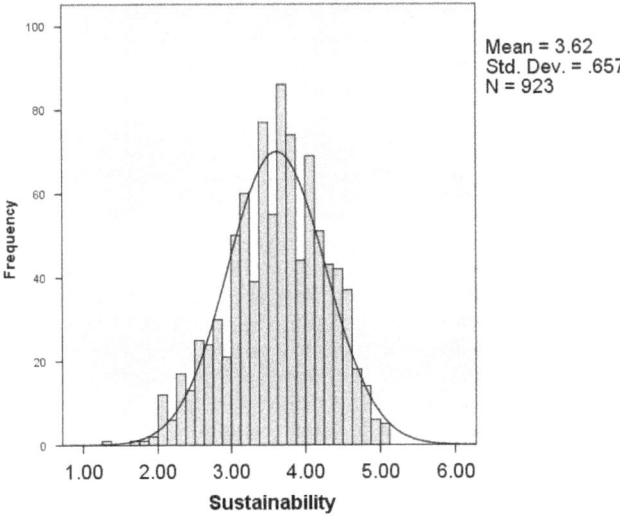

(caption on next page)

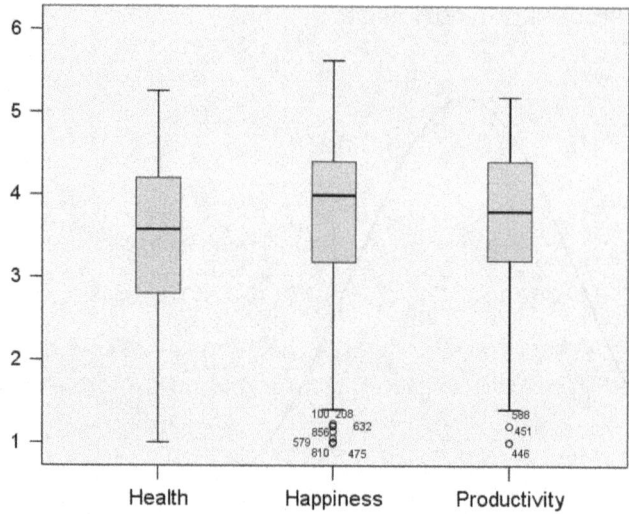

Figure 3.6 A representation of three distributions of interval level of measurement variables through boxplots – the picture represents distributions of scores of entities in the sample on three scales of career sustainability perception from an unpublished study by Hedrih and Milić. The vertical axis represents variable values (the scales of variables are comparable), the horizontal axis are the three variables and their names are listed. One boxplot is for each of the variables. The box part of the boxplot represents the range of variable values in which most of the population is grouped. The thick line accross the middle of each box is the mean of the sample in this case, although boxplots can be created with different measures of central tendency indicated. The "ropes" i.e. the thin lines going up and down from the box represent ranges of values that are rarer. Finally the numbers outside the values covered by the thin lines represent outliers. We can see on these boxplots that there are multiple outliers with unusually low values on Happiness and also that there are 3 outliers with very low values on Productivity (there are only two circles because 2 of the 3 outliers have the same value, so their circles overlap). The length of the boxplot and especially the length of the box indicate the variability of the group represented by the boxplot on the presented variable – the longer the boxplot and especially the longer the box, the higher the variability.

───

Figure 3.5 A representation of a distribution of an interval level of measurement variable through a histogram – the picture represents the distribution of the sample from an unpublished study (Hedrih & Milić) on the perceptions of career sustainability, which is one of the psychological constructs measured by the Career Sustainability Perception Scale developed by the authors of the study. This graphical depiction is created by converting the continuous variable into a set of intervals, a set of discrete values, so that it is possible to count the number of entities in each interval. The vertical axis represents the number of entities, which are study participants in this case and the height of each bar corresponds to the number of entities in the interval of variable values represented by the bar. A probability density curve is drawn over the bars, based on the distribution shape indicated by the relative sizes of bars of the histogram. We can see from this particular chart that most of the participants tend to have values in the central area, while the number of participants becomes smaller and smaller as values go away from these central areas of the distribution.

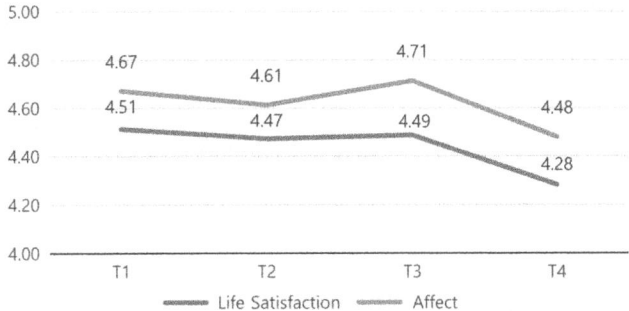

Figure 3.7 A line graph used to represent central tendencies of multiple samples on two different variables. The points on the horizontal axis are 4 different time points when measurements were taken, while the vertical axis contains variable values. Each line represents one variable. Points on the line above T1, T2, T3 and T4 and the numbers written above the lines represent the arithemtic means of the sample on the two variables at those timepoints. For example, we can see that at timepoint T3, the sample mean on the variable Affect was 4.71, while it was 4.49 on Life Satisfaction. At time point T1, the sample mean on Affect was 4.67 and 4.51 on Life satisfaction. The figure is from a paper by Kwon & Lee (2020). Reprinted by the permission of Elsievier, licence number 5132920194933.

Figure : Relative ratio of male to female COVID-19 infection rates (M/F ratio) per 1,000 population by age; ten European countries

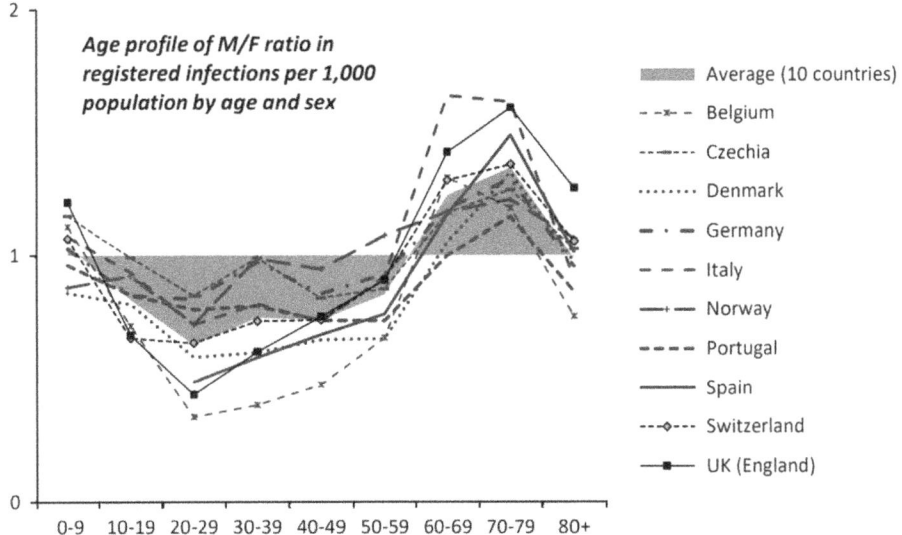

Figure 3.8 A more complex graphical representation of a comparison of multiple distributions. The graph represents the ratio of infected males and females in various countries depending on gender. Each line refers to a different country. The vertical axis represents the male-to-female ratio of COVID-19 infections, obtained by dividing the number of registered COVID-19 infections of males by the number of COVID-19 infections of females in the same age groups, while the horizontal axis represents different age groups in years. In the picture, we can see that, at the age range 20–29, there are more females than males reported with COVID infections (because the ratio is lower than 1, meaning the divisor is higher than the dividend in the ratio), while in the age groups of 60–69 and 70–79 there are much more males infected than females. The figure is from a preprint by Sobotka et al. (2020). Reprinted with the permission of authors.

3.6 Let us apply what we learned so far

Let us now try to apply what we have covered in this chapter through a couple of exercises. Please refer to the start of the book for the general instruction for completing the exercises. Our suggestion is that you first read the excerpt and the statements and provide your own answer. You may write it in the Answer column, and after that look up the answers and compare your own answers with them.

Exercise C. Distribution, measures of central tendency, measures of variability

Descriptive statistics.

		(n = 252)	(%)
Gender	Male	84	33.3
	Female	168	66.7
Age (years)	20-24	51	20.2
	25-29	106	42.1
	30-34	71	28.2
	35-39	24	9.5
Marital status	Married	29	11.5
	Single	223	88.5
Occupation	Employee	174	69.1
	Student	50	19.8
	Other	28	11.1
Education level	Less than high school degree	7	2.8
	Bachelor's degree	202	80.2
	Graduate degree	43	17.1
Income	Under $20,000	15	6.0
	$20,000 ~ $29,999	43	17.1
	$30,000 ~ $39,999	42	16.7
	$40,000 ~ $49,999	45	17.9
	$50,000 and over	107	42.5

Table is from: Kwon, J., & Lee, H. (2020). Why travel prolongs happiness: Longitudinal analysis using a latent growth model. Tourism Management, 76, 103944. https://doi.org/10.1016/j.tourman.2019.06.019. Reproduced by permission from Elsevier, license number 5132920194933.

C	Statement:	Answer
C1.	Numbers in the column on the right-hand side are percentages.	
C2.	Gender is operationalized as a binary variable in the paper.	
C3.	Most of the people in the sample have a graduate degree.	
C4.	There are over 200 females in the sample.	

C	Statement:						Answer
C5.	Most of the males in the sample are under the age of 29.						
C6.	There are more people in the sample earning $50,000 and over than people earning less than $29,999.						
C7.	More than 2/3 (two thirds) of the study participants are unemployed						
C8.	Marital status is operationalized in this study as a nominal variable with 4 categories.						
C9.	Median income for this sample is between $20,000 and $30,000.						
C10.	Absolute arithmetic mean of the sample is above the threshold median.						

Exercise D. Distribution, measures of central tendency, measures of variability

			Item number 20					Total
			1	2	3	4	5	
Gender	male	Count	50	20	22	12	9	113
		% within gender	44.2%	17.7%	19.5%	10.6%	8.0%	100%
		% within answer	41.0%	44.4%	53.7%	63.2%	69.2%	47.1%
		% of total	20.8%	8.3%	9.2%	5.0%	3.8%	47.1%
	female	Count	72	25	19	7	4	127
		% within gender	56.7%	19.7%	15.0%	5.5%	3.1%	100%
		% within answer	59.0%	55.6%	46.3%	36.8%	30.8%	52.9%
		% of total	30.0%	10.4%	7.9%	2.9%	1.7%	52.9%
Total sample		Count	122	45	41	19	13	240
		% within gender	50.8%	18.8%	17.1%	7.9%	5.4%	100%
		% within answer	100.0%	100.0%	100.0%	100%	100%	100%
		% of total	50.8%	18.8%	17.1%	7.9%	5.4%	100%

The table shows the distributions of answers of study participants (people) on the item number 20 in a questionnaire. The items were on a rating scale with possible answers between 1 and 5, with 1 meaning total disagreement with the statement contained in the item and 5 meaning total agreement. The table presents the distributions of the whole sample and on male and female subsamples separately, as well as percentages for the distribution by gender within each answer. Data are from a study carried out by authors of this book.

D	Statement:	Answer
D1.	There are more females who answered 1 on item 20 than males who gave the same answer.	
D2.	Of all the females in the sample 44.2% answered 1 on item 20.	
D3.	Of all the males in the sample, 17.7% answered 2 on item 20.	
D4.	Of all the people in the sample who answered 1 to item 20, 59% are females.	
D5.	Of all the females, less than 5% answered 5 on item 20.	

D	Statement:	Answer
D6.	There are more females than males in the sample.	
D7.	Mode of item 20 is 3 on the total sample.	
D8.	There are less than 200 people in the total sample.	
D9.	The number of females who answered 4 on item 20 is higher than the number of males who gave the same answer on this item.	
D10.	Mode (modal value) of the variable gender on this sample is female.	

Exercise E. Distribution, measures of central tendency, measures of variability

Mean micronuclei frequencies and nuclear division indexes in studied age groups. Total sample size = 133.

	Age range in years	MN (mean±SD)	NDI (mean±SD)
1	**0 (n = 29)**	0.56 ± 0.71	1.52 ± 0.30
2	**21–40 (n = 30)**	0.82 ± 0.78	1.72 ± 0.26
3	**41–60 (n = 29)**	1.26 ± 1.48	1.68 ± 0.35
4	**61–80 (n = 26)**	5.48 ± 3.65	1.55 ± 0.24
5	**81–92 n = 19)**	4.48 ± 2.69	1.62 ± 0.34

Legend: n – number of participants in a group. MN – micronuclei frequency per 1000 binuclear cells. NDI – nuclear division index. Means and standard deviations for each of the five age groups are presented.

Table based on data from authors' study published in Hedrih et al. (2018)

E	Statement:	Answer
E1.	Modal average is greater than 31 for all groups.	
E2.	The frequency of micronuclei (variable MN) is the highest for the group aged 61-80.	
E3.	The highest variability in the frequency of micronuclei (variable MN) is in the group of newborns (age 0).	
E4.	The age group with the highest mean on the variable MN has at the same time the highest mean on the variable NDI.	
E5.	The frequency of micronuclei (variable MN) is on the nominal level of measurement.	
E6.	The variance of the variable MN for the age group of newborns (age 0) is higher than 1.	
E7.	There is not a single person in the age group between 61–80 years in the sample with the MN value lower than 1.	
E8.	All newborns in the sample have NDI values below 1.5.	
E9.	On the variable MN, the two age groups below 40 years of age have lower variability than the age groups from 41 years of age and above.	
E10.	There were over 200 participants in the sample.	

Let us now consider the answers:

C1 − true. Those numbers indeed are percentages as indicated by the % sign in the column name.

C2 − true. We can see that there are only two categories reported for gender, meaning that it was operationalized as a binary variable.

C3 − false. We can see that only 43 people have a Graduate degree, while 202 or 80,2% have a Bachelor's degree.

C4 − false. We can see that there are 168 females in the sample and this is less than 200.

C5 − unknown. While we have both the number of males and the number of people below 29, we do not have a crosstabulation of these two variables, so we cannot tell from the data how many of the males are below 29.

C6 − true. We can see that 42.5% of people in the sample reported earning $50.000 and over, while the two categories below $29.999 contain 17.1% and 6.0% of the sample. Together these two categories are 23.1% of the sample which is less than 42.5%

C7 − false. We can see that the number of participants who listed their occupation as employee is 69.1%, which is more than two thirds of the sample. If we consider that people who listed their occupation as an employee must be employed, even if all the other study participants were unemployed, that would still be less than one third of the sample unemployed.

C8 − false. We can see from the table that there are only two categories of marital status − married and single.

C9 − false. To answer this question, we are looking for the median i.e the 50th percentile. If we sum up the percentages of categories of the variable Income, we would see that the categories up to $30,000 add up to less than 50%. Even if we added to it the category $30,000-$39,999 that would still not add up to 50%. For that we would need to add the category $40,000-$49,000, which means that the median income is in that category.

C10 − meaningless. There is no standard statistical concept called an "absolute arithmetic mean". The same goes for the "threshold median". If an author wanted to introduce such concepts, he/she would have to define them first.

D1 − true. We can see from the table that there are 72 females compared to 50 males who answered 1.

D2 - false. We can see that the % of females (% within gender) who answered 1 is 56,7%. The number 44,2% is the percentage of males.

D3 − true. We can see that % within gender for males on answer 2 indeed is 17,7%.

D4 − true. We can see that % within the answer for females is 59%. This means that of all the people who answered 1, 59% were females (while 41% were males).

D5 − true. We can see that only 4 females answered 5 and that this is 3.1% of all females.

D6 − true. We can see from the table that numbers are 127 females and 113 males.

D7 − false. We can see that the mode i.e. the most frequent answer on item 20 on the total sample is 1 and not 3. 122 or 50,8% of the total sample answered 1.

D8 − false. No, the total sample size is 240 as we can see from the total count.

D9 − false. We can see that 7 females answered 4, but 12 males. So, the number of males who gave this answer is higher, making this statement false.

D10 − true. We can see that gender in this table has two values and the frequency of females is higher, making them the mode − females are 52.9% of the sample, while males comprise 47.1%, making females the mode of the sample.

E1 − meaningless − There is no such thing as a "modal average".

E2 – true. Yes, we can see that the average frequency of micronuclei for this group is 5.48, which is higher than any other mean value for this column (compare this to all the other first numbers in the MN column).

E3 – false. We can see that the measure of variability – standard deviation in this case, for the group of newborns (age 0) is .71 and this is not the highest value in the MN column (i.e. for the MN variable). Actually, it is the lowest variability of all groups.

E4 – false. We can see that the highest variability on the variable MN is in the group 4, i.e. among people between 61 and 80 years of age, but that the highest standard deviation on the variable NDI is in the group between 41 and 60 years of age.

E5 – false. Since the authors calculated the mean and the standard deviation for it, this variable must be at least on the interval level of measurement, as this is the minimum required level for these statistics.

E6 – false. Variance is standard deviation squared (standard deviation multiplied by itself). Since we can see that the standard deviation for this group is .71 i.e. lower than 1, variance of this group must also be lower than 1, because squares of numbers lower than 1 are lower than that number, which means also lower than 1.

E7 – unknown. While we can see that the mean of this group is 5.48, which is a central tendency measure, we do not have a presentation of the whole distribution for this sample. That means that it is possible that there might be someone in the group with a value below 1, but we do not know. Also, given that the standard deviation for this group is 3.65, a value of 1 is certainly possible.

E8 – false. We can see that the mean of the group of newborns on the variable NDI is 1.52. Therefore, for this statement to be true, it would be necessary that all study participants from this group have values below the mean of their group and, knowing how the mean is calculated, we can conclude that this is impossible.

E9 – true. We can see that the standard deviations of the first two groups are both below 1, while the standard deviations of the three other groups are higher. Since the standard deviation is a measure of variability, the statement about the variabilities of the groups is true.

E10 – false. We can read in the upper part of the table that the sample size is 133, which is lower than 200.

References

Harding, B., Tremblay, C., & Cousineau, D. (2014). Standard Errors: A Review And Evaluation Of Standard Error Estimators Using Monte Carlo Simulations. *The Quantitative Methods for Psychology*, *10*(2), 107–123. 10.20982/tqmp.10.2.p107

Hedrih, Anđelka, & Hedrih, V. (2012). Attitudes and Motives Of Potential Sperm Donors in Serbia. *Vojnosanitetski Pregled*, *69*(1). 10.2298/VSP1201049H

Hedrih, Andjelka, Najman, S., Hedrih, V., & Milošević-Đorđević, O. (2018). Structure Of Relations Between The Frequency Of Micronuclei In Peripheral Blood Lymphocytes And Age, Gender, Smoking Habints And Socio-demographic Factors In South-east Region Of Serbia. *Facta Universitatis. Series Medicine and Biology.*, *20*(2), 47–54. 10.22190/FUMB180102008H

Hedrih, V. (Ed.). (2017). *Work and Family Relations at the Beginning of the 21st Century*. Filozofski fakultet, Niš.

Hedrih, V. (2020). *Adapting Psychological Tests and Measurement Instruments for Cross-Cultural Research: An Introduction (1st Edition)*. Routledge, Taylor&Francis Group.

Kwon, J., & Lee, H. (2020). Why Travel Prolongs Happiness: Longitudinal Analysis Using A Latent Growth Model. *Tourism Management*, *76*, 103944. 10.1016/j.tourman.2019.06.019

Sobotka, T., Brzozowska, Z., Muttarak, R., Zeman, K., & Lego, V. D. I.. (2020). Age, Gender and COVID-19 Infections. *MedRxiv*, 2020.05.24.20111765. 10.1101/2020.05.24.20111765

4 Distributions

4.1 Theoretical and empirical distributions

As stated in the previous chapter, a distribution is a listing or a function describing all the existing (or all possible) values of a variable and how often they occur. It is usually created by listing each specific value of the variable and the frequency with which it occurs, but it can alternatively be also presented by creating intervals of values for continuous variables or broader categories for variables with many different possible values (such as continuous variables) and then listing the frequencies of entities in each of these intervals or broader categories. It can also be presented as a probability density function.

Aside from providing a simple factual description of values of sample entities on considered variables, distributions are also used in statistics to make various inferences about the circumstances around properties of the variable, the sample and the measurement based on the shape of the distribution. This is done by comparing the distribution obtained from the empirical data to certain known shapes of distributions, distributions that are known to be the result of certain known processes. If the shape of the sample distribution resembles the shape it is compared with, we can infer, provided other properties of the sample and the measurement allow it, that processes known to result in such distributions likely took place. Or otherwise, we can conclude that our distribution likely was not the results of processes that result in distributions of a certain shape. For this reason, we will make a distinction between empirical and theoretical distributions:

- An **empirical distribution** is a real distribution we have obtained from real data i.e. from real observations or measurements. It can have any shape, depending on the data and the properties of the sample from which it was obtained.
- **Theoretical distributions** have a certain shape, a name and are based on a theory or a similar scientific vehicle that describes situations in which we can expect them. These theories provide insight into what type of distribution we expect with a certain type of data and are thereby the basis for predicting probabilities of certain events.

4.2 Normal distribution

The most widely used and the most well-known theoretical distribution is the **normal distribution,** sometimes also called the Gauss distribution or the Gauss-Laplace distribution.

DOI: 10.4324/9781003107712-4

It is a continuous distribution typically presented through its probability density function called the **normal curve, Gauss curve or the bell curve.** Normal distribution is symmetrical with the entities being the most frequent at its center and the frequencies than dropping more and more with distance from the center in both directions. It asymptotically approaches the frequency/probability of zero, meaning that with distance of a variable value from the center of the distribution, the probability of an entity having that value is falling more and more, but it never reaches zero, only becomes ever smaller and smaller into infinity.

Normal distribution is the expected distribution of data in a very wide range of situations that can be described as situations where the population values are influenced by one very influential factor that is a constant for all entities and a wide array of much less influential and mutually independent, random factors that vary between entities and that are the cause of differences between them. Such situations include individual differences between people or other organisms on various variables, achievements on educational tests (in situations when people of similar ability undertook the same educational program), psychological tests, results of repeated measurement of the same phenomenon using instruments of imperfect precision i.e. measurement errors, various indicators of market dynamics etc. The concept of the normal distribution has been known and in use for centuries and it is uncertain who pioneered the concept, although the first contribution to the development of the concept is often attributed to the 18th century French mathematician Abraham de Moivre and his book on probability published in English titled " The Doctrine of Chances: Or, A Method of Calculating the Probabilities of Events and Play" (de Moivre, 1756) (Figure 4.1).

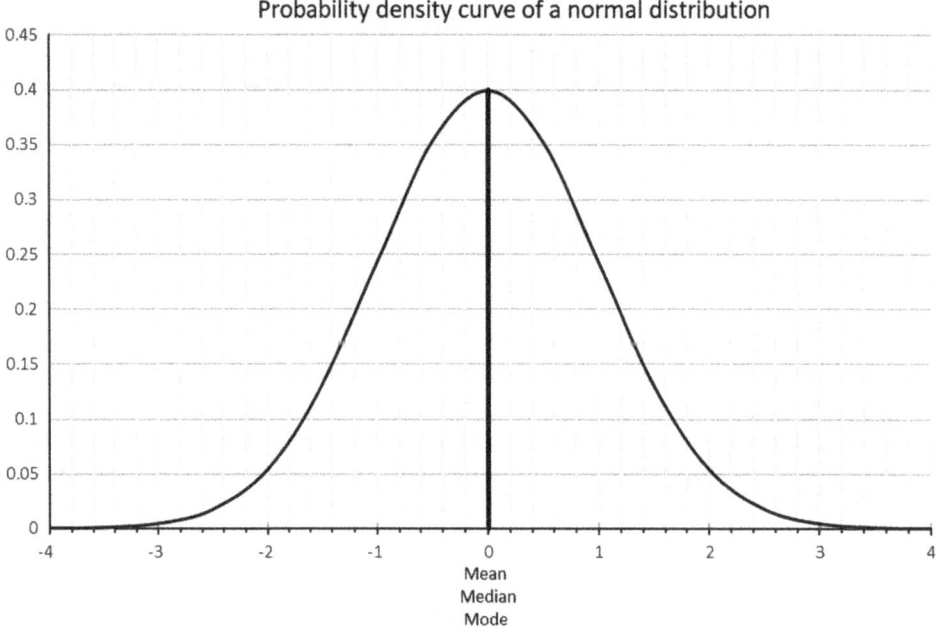

Figure 4.1 Representation of a normal distribution through a probability density function. The point marked by 0 is the arithmetic mean, the median and the mode, which all have the same value on an ideal normal distribution. The vertical axis represents probabilities, while the horizontal axis represents the values of the variable.

How is the concept of normal distribution used in practice? As stated earlier, a normal distribution is the expected distribution in situations when values of the variable are the result of one very influential factor that is a constant and a large number of independent random factors that are variables. Most commonly, that is the expected situation when we are dealing with individual differences between entities in a population with regard to a certain variable. We can use this assumption about how the normal distribution occurs to test whether it holds for a particular population we are examining on a particular variable. For example, we can examine whether individual differences in height in inhabitants of a certain area follow the normal distribution i.e. whether the individual differences are just due to various small random factors or whether there are some more important factors at play. We could than take a sample from that population, measure their heights and compare the empirical distribution thus obtained with the theoretical normal distribution. If it turned out they are a match, this would mean that we are looking at a homogenous population according to height. If it turned out that they are not a match, we could examine the differences and then make conclusions about what happened. What we would, in this particular example, likely find is that the biological sex is also a strong factor determining height and that if we wanted to have a normal distribution, we would have to study participants of the same sex. This would be due to the fact that body height is a sexually dimorphic trait in humans. However, depending on the population studied, if there are, within the population, groups that suffered from chronic malnutrition during childhood or were systematically influenced by some other factor affecting height, we would likely be able to see that in the shape of the height distribution, provided of course that the group affected by such a factor is sufficiently large. In a similar way, if we gave a test covering the program of a curriculum to a group of students, if they all undertook the same course, we could expect the distribution of their test results to be normal. If it were not normal it could point to some systematic factors behind the deviation from the normal distribution. For example, we could discover that there were some students who did not follow the course at all, but just showed up for the test, or it could sometimes point to a validity problem with the test. In a similar fashion, if we looked at an athlete training for a race and measuring the time it took him/her to run the course, we could also expect that the distribution of the needed time follows the shape of the normal distribution. If we found out that it did not, this could point to some systematic factors – was he/she becoming progressively more tired in subsequent runs? Did his/her motivation change? Were there some obstacles at the course present in certain runs? Maybe he/she was injured or hurting during certain runs etc.

****

If we take two independent, normally distributed variables and calculate their ratios i.e. we divide values of two variables on each entity with each other, these ratios would form a distribution called the **Couchy distribution,** named after the 18th–19th century French mathematician Augustin-Louis Couchy. The Couchy distribution is also referred to as the Lorenz distribution, the Cauchy-Lorenz distribution or the Breit-Wigner distribution.

4.3 Poisson distribution

The Poisson distribution describes the number of times a certain random event happened within a given unit of measurement - a time period or a unit of space. This random event is such that it has a fixed mean rate of occurrence in each unit of measurement. In this concept, the probability of the event happening in each unit of measurement is fixed and occurrences of the event are independent of each other. The unit of measurement can be a certain time period, a unit of space, a certain distance, or area, but can also be an individual person. In social sciences, the Poisson distribution is often used to describe the number of times every member of the population experienced a certain random event during a certain time period. The assumption is that the probability of this event happening per time unit is fixed and equal for all population members.

The Poisson distribution is a discrete distribution i.e. its values can only be natural numbers, since it represents the numbers of time an event occurred, which is always a natural number. It was named after the 18th–19th century French mathematician Siméon Denis Poisson (Figure 4.2).

How is the Poisson distribution used in practice? It is useful in a variety of situations where the goal is to determine whether certain events are due to random chance or if there are some systematic factors at play. For example, if we want to determine whether errors in industrial production happen because of chance or because, for example, certain workers fail to follow proper procedures we can start with the expectation that errors are simply due to the fact that production procedures are such that they necessarily result in a certain percentage of random errors. In that case, the mean number of errors during a fixed time period should be equal across workers doing the same job activities. This however, does not mean that after a certain period of time all workers would have the same number of errors, because they would not, It means that the distribution of the number of errors per worker would have the shape of a Poisson distribution, with some

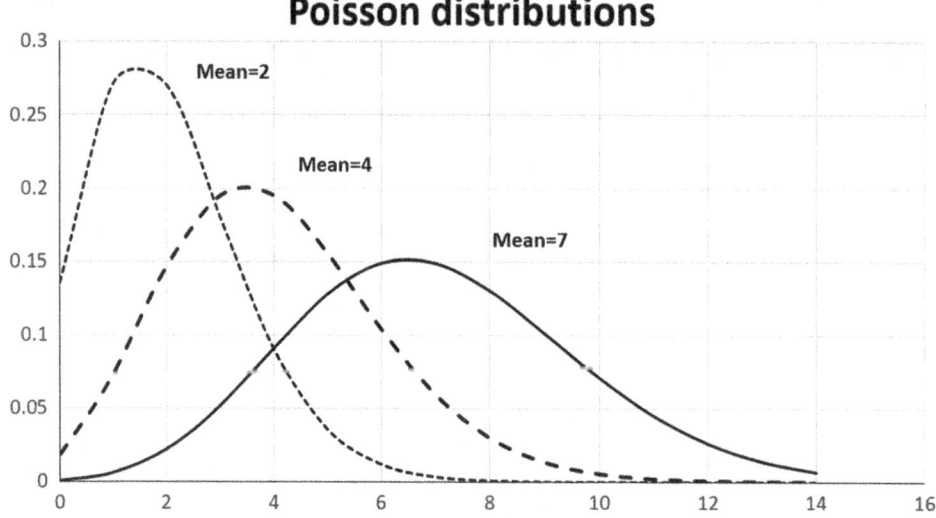

Figure 4.2 Shapes of the Poisson distribution depending on the mean of the distribution. When mean is high enough i.e. when the center of the distribution is sufficiently distant from 0, the shape of the Poisson distribution resembles the shape of the normal distribution. Poisson distribution is a discrete distribution and its values are counts of the number of times an event occurred per entity, therefore on the ratio level of measurements and can be only natural numbers. The vertical axis are probabilities, while the horizontal axis represents the number of times an event occurred (per entity). The distribution curve represents probabilities that a random entity from the population in which the distribution parameters hold will have a certain value on the horizontal axis.

workers having made more errors, some having made less errors, but the mean being equal to the basic expected mean error rate of the procedure. We could than compare the real distribution of the number of errors per worker to the Poisson distribution with the same mean rate of error occurrence and see if the two distributions match. If they are sufficiently similar, we could conclude that the errors are indeed the result of random chance. In that case, if we wanted to further reduce the occurrence of errors, we would have to rethink production procedures making them less prone to resulting in error. However, if we found that the two distributions do not match, but that there are profound differences, we would have to examine the nature of these differences. We could then, for example, discover that there are is a group of workers making much less errors than what we would expect based on the distribution. This could then mean that these workers might be using different procedures that are better than standard ones. We could then study what they are doing and adopt it elsewhere. Or we could discover that there are workers that make errors much more often than expected. This would mean that there is likely some unwanted systematic factor at play with that group and we would need to study what it is, expecting to find things such as insufficient or inadequate skills or training, lack of discipline or the ability to follow procedures or some other unwanted factor specific for that particular group. What the Poisson distribution thus enables us it to tell which differences in the number of occurrences of the event in focus (errors in production in the given example) might be due to random chance and are thus expected, and where we could meaningfully look for systematic factors at play.

In a similar way, the comparison of the empirical distribution with the Poisson distribution could help us determine things like, for example, are there hours in a call center where the rate of calls received is lower or higher than normal, identify when traffic accidents are due to random chance and when they might be due to certain routes being more dangerous or due to driver skill or care, when the number of patients arriving at a clinic is within normal random variation and when the difference might be due to some extraordinary factor etc.

4.4 Binomial distribution

The binomial distribution describes the number of times a random event occurred to each entity of a population in the scope of a certain predefined number of trials. In other words, it is a discrete probability distribution of the number of events occurring in a sequence of a certain predefined number of independent trials. The event in question has a certain fixed rate/probability of occurrence with each trial and we record in each trial whether the event happened or not. A trial referred to here is a specific activity that can result in the random event in question occurring or not occurring. The outcome of these trials is always binary (happened/did not happen) and the number of trials for each entity in the population (or the sample) is the same. These trials are called Bernoulli trials. The special case of the binomial distribution where the number of trials (per entity) is 1 is called the Bernoulli distribution.

It can be noted that the conditions under which a binomial distribution can be expected are similar to those for the Poisson distribution with the main difference being that the Poisson distribution is based around the units of measurement – time intervals, units of space etc. and that the event can occur multiple times inside that unit, while the binomial distribution is based on Bernoulli trials and the event can occur only once in a trial. The distribution is named after the 17th century Swiss mathematician Jacob Bernoulli (Figure 4.3).

Binomial distributions
15 trials

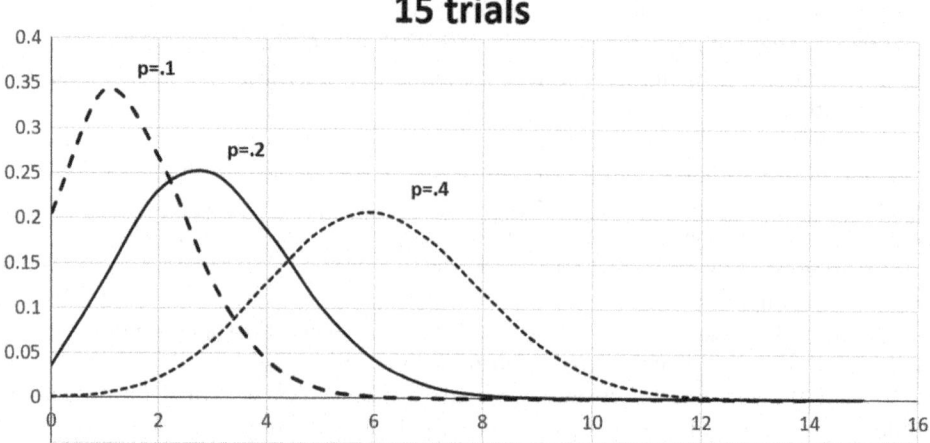

Figure 4.3 Examples of shapes of the binomial distribution with 15 Bernoulli trials and with different probabilities of the event occurring during a trial. When the center of the binomial distribution is sufficiently distant from 0, the shape of the binomial distribution resembles the shape of the normal distribution. Numbers above the curves indicate the probability of the event occurring in that distribution described by that particular curve. Binomial distribution is a discrete distribution and its values are natural numbers that cannot be higher than the total number of trials i.e. the maximum of the binomial distribution is when the event happens in every trial. The vertical axis are probabilities, while the horizontal axis represents the number of times an event occurred (per entity). The distribution curve represents probabilities that a random entity from the population in which the distribution parameters hold will have a certain value on the horizontal axis.

The practical use of the binomial distribution is similar to how the Poisson distribution can be used with the only difference being that the situations need to involve processes that can be adequately represented by Bernoulli trials i.e. processes with binary outcomes that are of interest to the researcher. For example, in a railroad company we could compare the number of late arrivals of each driver after a certain number of drives along a standard route to the binomial distribution to make inferences about whether these late arrivals are likely due to random chance and the fact that there are simply random factors that equally affect all drivers and that could cause them to be late or if there are some systematic factors at play with certain drivers (e.g. insufficient train driving skills, bad time management…) or trains (e.g. technical defects). We could also, for example, use the binomial distribution as a reference to gain insight whether the current success rate of an athlete practicing pole jumping (pole vaulting) is in line with his/her regular performance or whether it might be improving or deteriorating due to some unknown factors etc.

4.5 Uniform distribution

A uniform distribution describes a situation where all variable values have the same frequency i.e. when probabilities of all categories/values of the variable are equal. It is most often used with data on the nominal level of measurement, but there are also special situations when it can meaningfully be used on data on higher levels of measurement. It also requires that the data be discrete, best if grouped into a small number of broad categories.

The uniform distribution is the most commonly used in situations when there is a need to determine whether a sample comes from a population where all values of the considered variable have equal probabilities i.e. where their frequencies are the same. Such situations include testing various pseudorandom number generators such as those used in games of chance, but also various research situations when there is a need to determine whether all categories are equally probable. For example, we can test whether a dice (a small cube with numbered sides for playing games of chance, that can be regarded as a simple form of a pseudorandom number generator) works correctly by playing it, for example, several hundred times and then comparing the obtained distribution of outcomes with the uniform distribution. If they are similar enough, the dice works correctly, if they are not similar enough, we could conclude that we have a so-called "loaded dice" i.e. a dice that more often lands on a certain side than on others. A correctly working dice should be landing on each side equally often. In the same way, we can ask ourselves whether two or more different products are equally often bought by the customers or whether the customers actually prefer a certain product over the others. We could test this by comparing the empirical distribution of products bought over a period of time, the distribution stating which product was bought how many times during the examined period, to the uniform distribution. If the two distributions are similar enough, we could conclude that customers did not exhibit a preference towards any of the products and that they buy all of them equally. If the two distributions are not sufficiently similar, we could conclude that customers have a higher preference for one or more of the examined products and could then inspect the empirical distribution to see which products are bought more often than others.

4.6 Deviations from the normal distribution

The normal distribution is likely the most frequently used type of theoretical distribution in research. It is the default expected distribution of individual differences on a variety of variables and, due to this, the precise properties of the deviation of the distribution of real data from it can often allow us to make various inferences about the data. That is the reason why researchers not only assess whether their data resembles the normal distribution or not, but are also interested in describing these deviations. As said earlier, the theoretical normal distribution is symmetrical and its curve follows a certain shape. The ends of the normal distribution asymptotically approach zero and, because the distribution is symmetric with the highest concentration of entities in the middle, its arithmetic mean, median and mode are equal (have the same value) and are located at precisely the center of the distribution. Empirical data can deviate from this either vertically or horizontally or in various more complex ways.

Horizontal deviations from the normal distribution happen when the empirical distribution is **asymmetrical** i.e. when, looking at the distribution, we can notice that entities tend to be grouped more tightly on one end and more spread out on the other end of the distribution. In a graphical representation, this will be present like one side of the distribution being shorter than normal and the other side being elongated. These asymmetric distribution shapes are referred to as positive or negative depending on whether the elongated part of the distribution is above the center of the distribution or below it i.e. whether the difference between entities on the elongated part of the distribution from the mean is positive or negative. The normal distribution that is not asymmetrical is referred to as a **symmetrical distribution**. Therefore, regarding the type of asymmetricity, a distribution can be:

- A **positively asymmetrical distribution** is a distribution shape in which the elongated side is the upper end of the distribution, the one in the area of higher, or positive values of the variable. Entities are concentrated in the lower part of the distribution, while the upper part of the distribution, the part with values higher than the center of the distribution is elongated. In a positively asymmetric distribution, **the mode is lower than the median** and **the median is lower than the mean**. A positively asymmetrical distribution usually indicates the presence of factors that prevent entities from attaining higher values. What such factors are exactly, of course, depends on the area of science. For example, in psychological or educational testing, a distribution like this typically indicates a difficult test i.e. a test consisting of items whose difficulty is above the level of the measured property of most participants in the sample. If a distribution like this is obtained on a test of knowledge, it would typically indicate that most test takers (from the sample) had insufficient knowledge on the topic of the test. If a cognitive test or a test of physical abilities or a set of physical exercises is in question a distribution like this typically indicates that test tasks were too difficult and above the level of ability or skill for most test takers. If an attitude or another type of conative test is in question it would indicate that attitudes or preferences of test takers are very different from those implied by test items. Unless the idea behind testing of this type was to be able to finely differentiate entities in the upper part of the distribution, those with highest values, a positively asymmetrical distribution is not a good development, as it means that differences in variable values (i.e. in test scores in case of tests) between entities in the center of the distribution will be reduced. As most of the entities are in the center that means that individual variability will be reduced between most entities. Given that we generally measure or assess things in order to be able to differentiate between them, this development goes directly against this goal and is, therefore, not good (Figure 4.4).

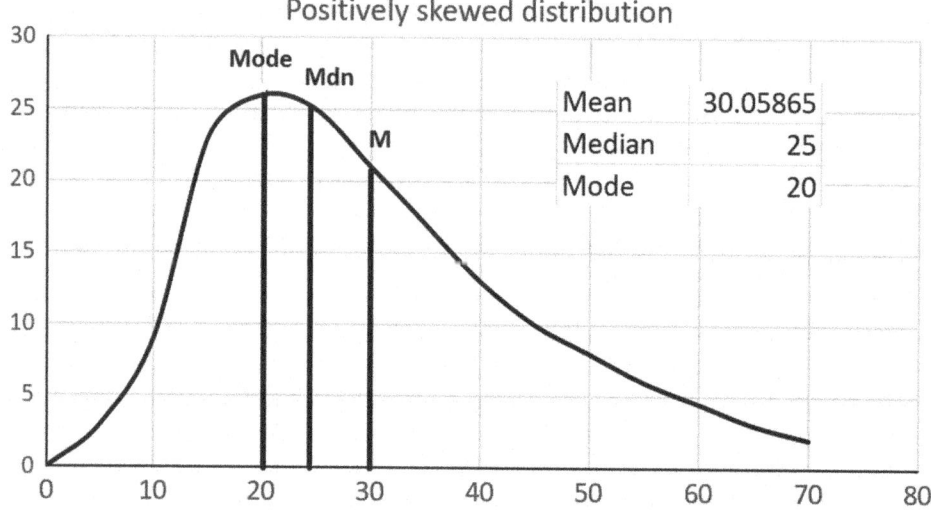

Figure 4.4 A graphical representation of a positively asymmetrical distribution. We can see from the picture that the mean is higher than the median and that the median is higher than the mode. The elongated, thick tail of the distribution extends towards the positive values, while the distribution ends abruptly towards the negative end. Numbers on the vertical axis are frequencies, while the numbers on the horizontal axis are variable values.

• A **negatively asymmetric distribution** is a distribution shape in which the elongated side is the lower part of the distribution. With this distribution shape, entities are concentrated in the upper part of the distribution, while they are spaced further apart and their frequencies are lower in the lower part of the distribution. Here, **mode is higher than the median** and the **median is higher than the mean**. In testing, a negatively asymmetric distribution usually indicates that the test was too easy. The general idea of testing is that some test-takers are able to pass just a few test items, some will be able to pass most test items and some will pass a moderate number of test items. In this way, test-takers with different levels of the variable assessed by the test will have different test scores. In contrast, a negatively asymmetric distribution indicates that most test-takers have high test scores and therefore, instead of their scores being spread out across the full span of possible test scores, high number of them is concentrated in a small interval of (high) test scores thus making differentiation of the levels of expression of the measured variable with them more difficult. For example, if we have a group of students in which almost all of them have the maximum grade, we will not be able to differentiate between their proficiency in the subject of that course (as we use grades for that and they all have the same grade – the maximum one) and that is as bad as everyone having bad grades, as far as the goal of differentiation between their achievements is concerned. Of course, from the standpoint of differentiation of achievements, a negatively and a positively asymmetric distributions are equally bad, with the only difference that one makes difficult the differentiation of the lowest achieving students and the other one of the highest achieving students (Figure 4.5).

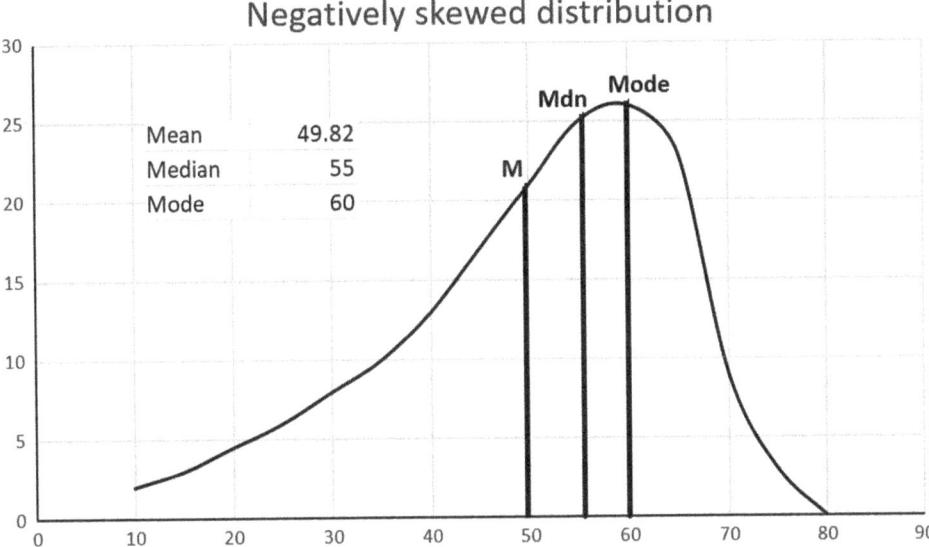

Figure 4.5 A graphical representation of a negatively asymmetrical distribution. We can see from the picture that the mean is lower than the median and that the median is lower than the mode. The elongated, thick tail of the distribution extends towards the positive values, while the distribution ends abruptly towards the negative end. Numbers on the vertical axis are frequencies, while the numbers on the horizontal axis are variable values.

The statistic showing the level of horizontal deviation from the normal distribution is called **skewness**. Different authors have proposed different ways of calculating skewness, but with most formulas, skewness calculation is **based on the relationship between mean and another measure of central tendency, on the number of entities above and below the mean or on the size of positive and negative differences between values of entities and the mean** in the two types of asymmetric distributions and also on the fact that **the more asymmetrical a distribution is, the larger these differences will be**. For example, one relatively simple formula for skewness based on the **comparison between the mean and the median** is:

Skewness = (3 * (Mean − Median))/Standard Deviation

Therefore, when the mean is higher than the median, this skewness statistic is positive, thus indicating that the distribution is positively asymmetric. When the mean is lower than the median, this skewness statistic is negative indicating that the distribution is negatively asymmetric. Of course, when skewness is zero (or sufficiently close to zero) the distribution is symmetric. Coefficient calculated by this formula is called the **second Pearsons coefficient of skewness** or the **Pearson median coefficient of skewness**. Another widely used skewness coefficient, called Pearson's moment coefficient of skewness is based on the differences between individual values from the mean (raised to the third degree). There is also a coefficient based on the difference between the mean and the mode (Pearson's first coefficient of skewness) and there are many others proposed by different authors. However, their basic interpretation is more or less the same – positive skewness means that the distribution is positively asymmetric, negative skewness means that the distribution is negatively asymmetric.

Another question that arises is **how to interpret skewness size**? How asymmetric does a distribution need to be to conclude that it cannot be considered normal? At the moment this book is written there is no definite answer to this question and various authors follow and propose different guidelines for interpreting skewness size. For example, certain authors (e.g. Field, 2009; Tošković, 2020) propose that skewness be interpreted by dividing the skewness coefficient with another statistic called the standard error of skewness (to be explained in the chapter about inferential statistics) and if the resulting value is above 1.96 or below −1.96 (or more liberally, 2.58 and −2.58) the distribution should be considered as non-normal, while if the statistic is within this range the distribution could be considered as not too asymmetrical. We can simplify this procedure somewhat by rounding 1.96 to 2 and phrasing it that this would mean that **skewness coefficients that are greater than double their standard error (greater than standard error*2) describe distributions that are not normal**. This procedure in a way assesses the probability that a distribution with skewness like the one to be interpreted can be obtained on a sample from a normally distributed population. However, the standard error statistic is a function of the number of entities in the sample and with very large samples it becomes very small. Then, dividing skewness with a number that is very small gives a result that is large and due to this, as the sample size increases, this procedure becomes less and less useful as the skewness size that passes the critical value becomes smaller and smaller, until a point where, with very large samples, even very minor deviations from the symmetrical shape exceed the boundaries. Some other authors therefore, propose various and somewhat arbitrary critical levels of various strictness. For example, Hahs-Vaughn & Lomax (2020) list **skewness of +3 and −3 as a very liberal range for distributions considered not too asymmetrical, that a moderate range would be between +2 and −2 and a conservative one would be between +1 to −1.**

Vertical deviations from the normal distribution happen when the distribution is "too pointy" with too many entities concentrated in the center and not spread out enough, or when the entities are too spread out, more dispersed around the mean than is the case with the normal distribution. Both of these situations typically happen when the premises the normal distribution is based on are compromised. For example, if the main factor that is supposed to be a constant and that determines the center of the distribution is actually not a constant, but has values that are sufficiently close for the distribution to still look like a single continuum (instead of two groupings of values) this might manifest itself as a distribution that is more spread out than the normal distribution. On the other hand, when there are too few factors aside from the main one to allow for variance or when there are factors restricting variability the result may be a distribution that is "pointier" than the normal distribution, with values of most entities more concentrated than is the case with the normal distribution. For example, in educational or psychological testing, when a test is too easy or too hard resulting in almost all the participants being grouped near the maximum or the minimum of the test, aside from being asymmetric, such a distribution will also likely be very "pointy" i.e. have values of test-takers more concentrated than we would expect from a normal distribution. The same might happen in situations when the students are doing a test together, cooperating or copying from each other, but submitting individual tests. Therefore, the variability of their answers will be lower than normal, even in cases when there were many incorrect answers, resulting in a concentrated, "pointy" distribution. On the other hand, when, for example, a society starts to get polarized around some political issue, with the population starting to get divided into two very different groups with regard to the attitude about the issue, the population will likely pass through a stage when the distribution of their attitudes about the issue will have a squashed, spread-out shape at a certain time point, before, if the polarization continues, splitting into two distinct groups.

In statistical literature, we will often see terms **platykurtic distribution** and **leptokurtic distribution** associated with these vertical deviations from the normal distribution. In contrast to these the normal distribution is called a **mesokurtic distribution.** Most commonly, the term **platykurtic distribution tends to be associated with "squashed" distributions**, where entities tend to be more spaced apart, while the term **leptokurtic distribution is commonly associated with the "pointy" distributions,** where entities are more concentrated than is the case with the normal distribution (Figure 4.6).

However, **such associations may not always be correct** as **there is much more to platykurtic and leptokurtic distributions than this**. Namely, the statistic used to

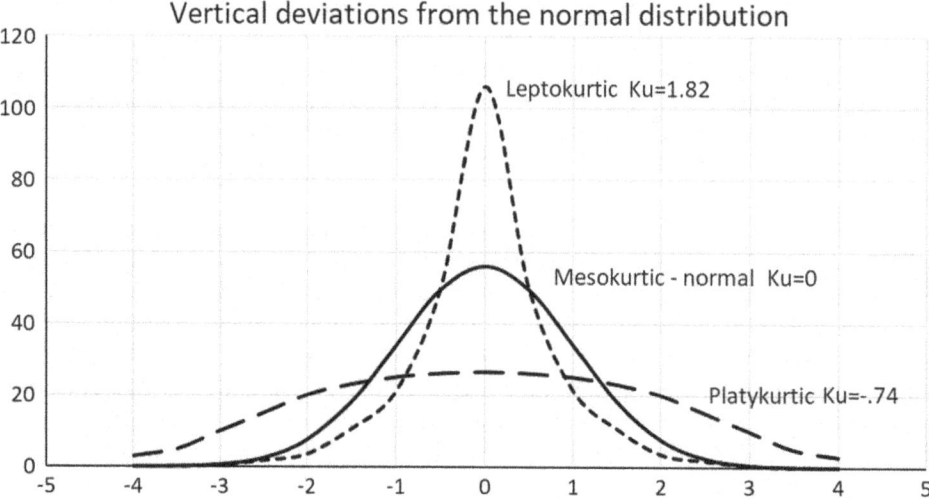

Figure 4.6 A graphical representation of vertical deviations from the normal distribution. We can see from the picture that the "pointy", leptokurtic distribution has a positive value of excess kurtosis, that the normal distribution has a zero value and that the excess kurtosis value of the "squashed", platykurtic distribution is negative. We should however bear in mind, that adding small numbers of outliers, entities to the ends of a normal distribution (to the tail) will increase kurtosis and that adding cases to the places on the distribution moderately away from the mean (to the shoulders) will tend to decrease it.

determine whether a distribution is platykurtic, leptokurtic or mesokurtic is called **kurtosis. Kurtosis formula is based on the average difference between individual values of entities and the mean raised to the fourth degree divided by the standard deviation raised to the 4th degree also**. It looks like this:

$$Ku = \frac{\frac{\sum_{i=1}^{n}(X_i - Mean)^4}{N}}{SD^4} - 3$$

In this formula, X refers to individual values of entities in the sample, mean is the sample mean and SD is the standard deviation. This part of the formula without −3 in the end is referred to as the **fourth standardized moment, and with −3 added to the formula**, the statistic obtained this way is called the **excess kurtosis**. If the **value of kurtosis**, calculated in this way, is **above zero i.e. positive, the distribution is leptokurtic**. If **the value of kurtosis is negative, the distribution is platykurtic**. Kurtosis value of a mesokurtic distribution is 0, in the presented − excess kurtosis formula. In the regular fourth standardized moment formula, the one without −3 added to it, value of the mesokurtic distribution is 3. The excess kurtosis formula corrects this to 0, by subtracting the number 3 from the fourth moment formula, to make the result more convenient for interpretation i.e to allow for the sign of the statistic to indicate the direction of the deviation from the normal distribution.

We should note that raising a difference from the mean to the fourth degree erases the sign, so both positive and negative differences add to the size of kurtosis. We should also note that for the same reason large deviations from the mean i.e. extreme values influence the total kurtosis value much more than small deviations. Therefore it is important to note that **kurtosis cannot be really treated as a measure of "pointiness" of the central part of the distribution**, as it is also **heavily influenced by the size of the ends or "tails" of the distribution**. (e.g. Darlington, 1970; Westfall, 2014).

If we look at the formula for kurtosis and examine how it works, we can conclude the following things:

- **adding entities around the center of the distribution will increase the value of kurtosis**, because it will reduce the standard deviation of the sample (which is a divisor in the formula for kurtosis). This is because these entities increase the sample size, and thus the divisor in the formula for the standard deviation, while increasing the sum of deviations from the mean much less because they are close to the mean, making their differences from the mean small.

- **adding small numbers of entities to the far ends of the distribution, particularly adding outliers will also increase kurtosis**, as they will increase the average of differences of entities from the mean raised to the 4th degree more than they will increase the standard deviation of the sample (even when raised to the 4th degree). This addition of outliers increases the value of kurtosis only as long as there are many times more entities closer to the mean than the number of added outliers i.e. as long as the number of entities in the sample is sufficient to make the influence of these added outliers on the standard deviation small. **Add too many outliers and they will start decreasing kurtosis instead of increasing it.**

- **If our variable was a constant**, i.e. if all entities had the same value, their standard deviation would be zero and this would **make kurtosis infinite or incalculable** as it would **include a division of zero by zero (0/0).** If we only had one or a couple of entities with a different value than the rest, kurtosis values would be very high instead of infinite/incalculable. However, **the minimum possible value of excess kurtosis, which is -2, can be obtained when there are only 2 different values of the variable and half of the entities have one, while half has the other value**.

In statistical literature, **the names that are common for these various parts of a normal distribution** (or a distribution similar to normal) are:

- **center** – the area around the central part of the distribution
- **tails** – the far ends of the distribution, both on the left and on the right, i.e. both on the upper and on the lower side considering the values of the variable.
- **shoulders** – the parts of the distribution between the tail and the center of the distribution. (Figure 4.7).

These **properties of kurtosis imply that distributions that are shaped very differently may have the same value of kurtosis**, thus **leading researchers to wrong conclusions**. For example, a mesokurtic or even a somewhat platykurtic distribution with a relatively small number of outliers might land a higher kurtosis value than a genuinely pointy distribution (i.e. with entities highly concentrated in the middle). A clearly "pointy" distribution with a substantial number of outliers added might have a 0 kurtosis, thus making the researchers believe that they are dealing with a normal, mesokurtic distribution. And there are other possible combinations, the common point with all of them being that they

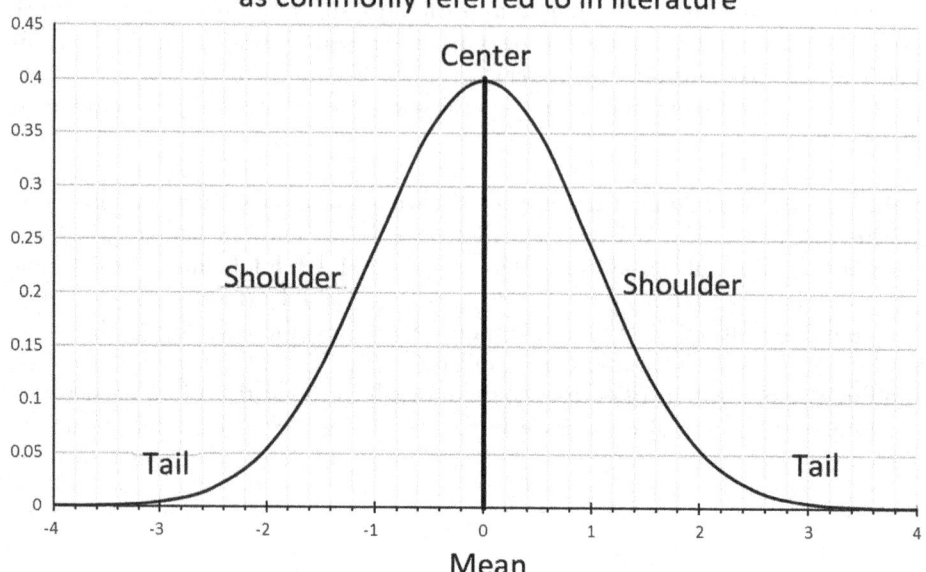

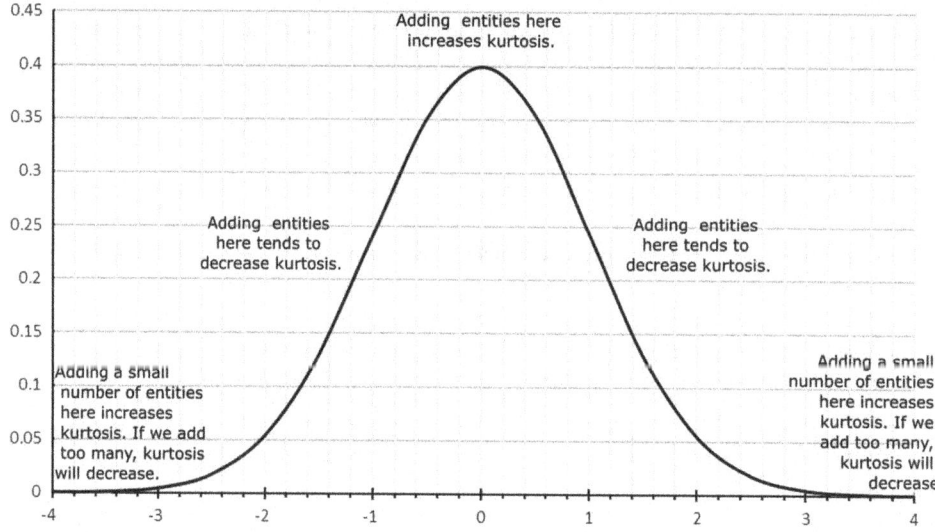

Figure 4.7 Parts of the normal distribution, the way they are typically referred to in literature (picture up). The center part of the distribution is referred to as the center, the ends of the distribution on both sides are called tails and the area between these is called shoulders of the distribution. The bottom picture provides an orientation about how kurtosis changes when entities are added to various parts of the normal distribution (picture down). Adding entities with values close to the mean will increase kurtosis. Adding entities with values that place them on the end of the distribution to the normal distribution will increase kurtosis up to a point, after which kurtosis will start to decrease with additional entities. Adding entities with values corresponding to the shoulders of the distribution will generally tend to decrease kurtosis. These effects refer to normally distributed samples. If the distribution of the sample is too different from normal, these effects may differ. Data on the vertical axis are frequencies, while the horizontal axis represents variable values.

would lead researchers to the wrong conclusions about the structure of the data, as demonstrated by examples presented in figures if not detected. (Figures 4.8 and 4.9).

Having this in mind, we should also point out that, usually, the primary and, most of the time, the only reason for calculating kurtosis is to find out whether our sample is normally

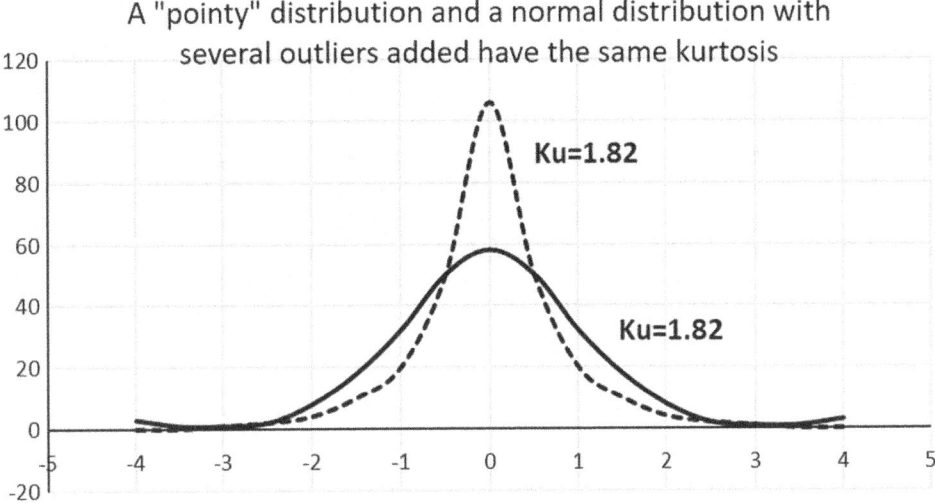

Figure 4.8 A graphical comparison between a normal distribution and a "pointy", high kurtosis distribution, to which a significant number of entities was added at the tails, thus reducing its kurtosis value to 0. Both distributions were initially created on samples of 280 entities, but additional 110 entities were added to the tails of the "pointy" distribution to reduce its kurtosis value to 0. This example shows that quite a significant number of entities (more than one third of the initial sample size) needed to be added to the sides of the "pointy" distribution to reduce its kurtosis to 0, turning this distribution into what is essentially a distribution with three focal points. It also shows how distribution shapes very much different from the normal distribution can still have a kurtosis value of 0 i.e. identical to that of the normal distribution. Numbers on the vertical axis are frequencies, while the horizontal axis represents variable values.

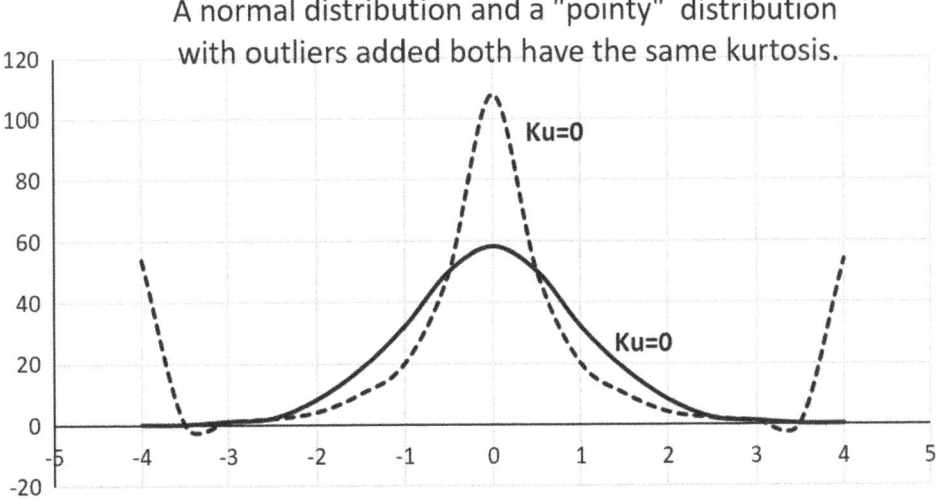

(caption on next page)

distributed or not. And when it is not, we are generally interested in finding out through the use of kurtosis whether it is "pointy" or "squashed". The distributions we are actually testing for are the one in which entities are very densely packed, highly concentrated and the one where they are more spaced apart than normal. In such a setup, **obtaining a high, positive kurtosis, on an otherwise normal distribution just due to the existence of a few outliers or obtaining a zero kurtosis for a highly pointy distribution with lots of outliers is clearly a very, very undesirable outcome**. This is even more so if we have in mind that outliers can often be a consequence of invalid measurement (e.g. people not doing the test seriously, or knowing the answers in advance) or even errors in data entry (e.g. entering 55, instead of 5, or entering 7 instead of 4, because it is the key above on the numeric keyboard!). Due to all this, **before interpreting the value of kurtosis, it is important to visually inspect the shape of the distribution and determine whether and to what extent are the values of kurtosis obtained influenced by outliers**. It might also be useful to calculate the value of kurtosis with a small number of cases trimmed from each end and compare the results. However, this trimming should not be taken to replace the visual inspection of the distribution, as many much more irregular distribution shapes than those discussed here may still result in normal kurtosis (and also skewness!) values. For example, a distribution with three distinct peaks might very easily have a skewness of 0 (this is essentially what figure 4.9. represents to a degree). Having all these issues in mind, we can likely conclude that, in spite of its popularity and widespread use, kurtosis may not be a particularly good indicator of the level of vertical deviation of a distribution from normal.

One **other thing that needs to be addressed** is the claim often found among researchers and even in literature that kurtosis is "a measure of thickness of distribution tails". This claim can be found in many scientific and educational texts, including textbooks in statistics. However, **kurtosis is not a measure of the thickness of tails of the distribution!!** If it were a measure of tail thickness, all high kurtosis distributions would need to have thick tails. However, this is not the case, and this is visible even in many texts that make such claim, who than proceed to present a leptokurtic distribution with a picture of a distribution with very thin or almost no tails. Although, as was discussed earlier, adding entities to the tails of the distribution i.e. making them thicker, can increase kurtosis, this is only up to a very close point. Adding more than a small number of entities (small relative to the sample size!) to the tails will lead to kurtosis starting to decrease. This means that there is an inverted U relationship between the number of entities added to the tails of the distribution and the change of kurtosis. An indicator whose direction of change inverts at a point with the increase in the property it is intended to indicate and which can also have high values, when the property it is intended to indicate is very low or 0 is no indicator at all. Also, we should keep in mind that the fact that the value of kurtosis is influenced by outliers and entities at the far in this way is, in most cases, an undesirable property of the kurtosis formula and not something it is intended to measure. That is why it is important to

Figure 4.9 A graphical comparison between a "pointy distribution" and a normal distribution to which several entities with values at the two ends of the distribution have been added and that now have the same kurtosis. Both distributions were created on samples of 280 entities and before the addition, the one with the lower center had the excess kurtosis value of 0, making it mesokurtic i.e. normal while the "pointy" one had the kurtosis of 1.82. Adding just 8 cases to the ends of the normal distribution led to their kurtosis values being equal, i.e. both distributions were leptokurtic. If a few more cases were added, kurtosis value of the distribution with the lower center would exceed the kurtosis value of the "pointy" distribution. However, adding even more cases to the ends, would quickly lower the kurtosis value again (see picture 4.7.). Numbers on the vertical axis are frequencies, while the horizontal axis represents variable values.

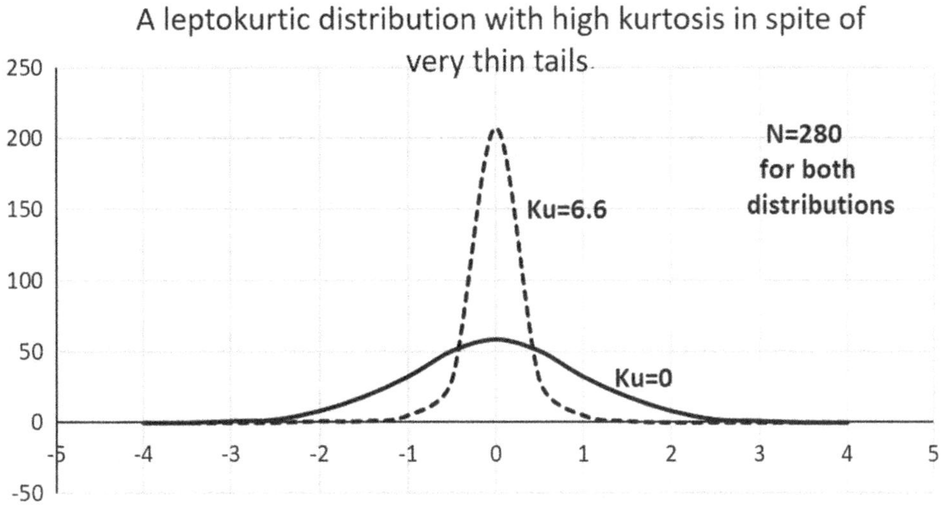

A leptokurtic distribution with high kurtosis in spite of very thin tails.

Figure 4.10 Kurtosis is not an indicator of the thickness of distribution tails! This example shows a distribution with almost no tails that, in spite of that, has a very high kurtosis value compared to a normal distribution. This goes counter to claims often heard from researchers and sometimes found in literature that kurtosis is an indicator of thickness of tails of a distribution. It is not, although thickness of the distribution tails i.e. the number of entities at the ends of the distribution do influence the kurtosis value in a complex way that has been described in this chapter. Numbers on the vertical axis are frequencies, while the horizontal axis represents variable values.

note **that claims describing kurtosis as an indicator of distribution tail thickness are simply wrong** (Figure 4.10).

After we have inspected the shape of the distribution and concluded that kurtosis value is not caused by some undesirable effect, i.e. that it can be regarded as a valid indicator of vertical deviation from the normal distribution, **how do we interpret the size of kurtosis?** As with skewness, there is no universally accepted method of interpreting the size of kurtosis. Like with skewness, one way to interpret it is to **divide the value of kurtosis with the standard error of kurtosis** (explained in the chapter on inferential statistics) and then to check **whether that value is between −1.96 and +1.96 (or between −2.58 and +2.58)** (e.g. Field, 2009; Tošković, 2020). This can also be simplified by rounding 1.96 to 2 and **stating that a distribution is not normal if its kurtosis value is more than twice as large as the value of its standard error of kurtosis**. Again, this method of assessment works appropriately with small samples, but not with large ones, as the standard error is a function of sample size (larger sample − smaller standard error). Due to this, some authors propose **that values of (excess) kurtosis of +1 and −1 be taken as boundaries**, with distributions whose kurtosis is inside the range being considered sufficiently close to normal, while those outside can be considered not normal (e.g. Hair et al., 2016). As noted earlier, **the value of (excess) kurtosis cannot be lower than −2.**

****

Aside from the described horizontal and vertical deviations from the normal distribution, empirical distributions can deviate from the normal distribution in various more complex ways. One such way is a situation when an empirical distribution has **more than one mode**. Such distributions are called **bimodal,** when they have **two modes** and

polymodal, when they have **more than two modes.** They are also sometimes referred to by the exact number of modes, such as threemodal, quartymodal etc. Bimodal and poly-modal distributions typically happen when the assumption of one single main factor that is constant for the whole population is compromised. This can also be taken to refer to situations when the sample is taken from 2 or more different populations. For example, one distribution often expected to be bimodal is the distribution of heights in the human po-pulation, with there being one mode for females and another one for males. If we look at salaries of employees in a certain group of widely existing profession (for example, waiters, or retail salesmen) but create a sample that consists of people taken jointly from one highly developed area with a strong economy and another very poorly developed with a struggling economy, their distribution will likely have two modes – one for the people from the poor area and another for the people from the wealthy area. In situations of intense political strife or a civil war, if attitudes of people from the opposing groups towards the main contested political issue are measured, it will likely result in a distribution with two or more modes (each group will have a separate mode). In situations like that, actually noting that the distribution has transformed to bimodal or multimodal can be an early sign that serious political strife is "around the corner" and might even indicate a threat of political violence or even a civil war within the society if the issue about which polarization occurred is sufficiently important and its potential adverse resolution seen as threatening to well-being or identity of at least one of the groups (Hamburger et al., 2021).

Polymodality of a distribution can best be detected by inspecting a graphical or tabular representation of the distribution and simply noting the number of peaks of the distribution and its general shape. A very low kurtosis value might also be an indicator of a polymodal distribution. However, it should be noted that **polymodal distributions might some-times have quite inconspicuous values of both skewness and kurtosis**, so **researchers are advised to always conduct visual inspection of the graphical representations of distributions of their data and not rely solely on the values of skewness and kurtosis** (Figure 4.11).

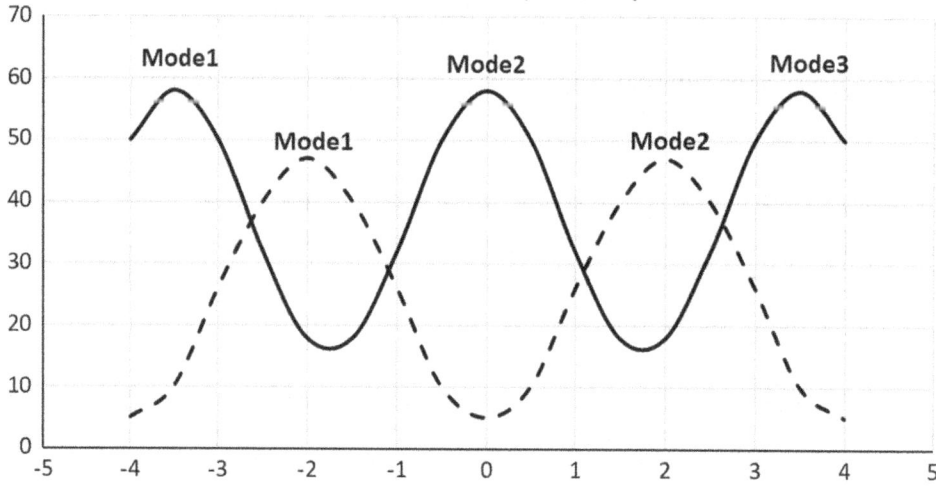

Figure 4.11 Examples of a bimodal distribution (dotted line) and a polymodal distribution with three modes (full line).

4.7 Standard scores and standardization

In research practice there is often a need to compare positions on a variable, either of two entities on the same variable or the same entity on different variables. While there is a variety of ways to compare two entities on the same variable, just comparing their raw values (values such as they initially are) does not typically directly translate into a position on the distribution. The issue becomes even harder when positions on distributions of two different variables need to be compared if, for example, these variables are on different scales and using different measurement units. One such question, requiring the comparison of variables on different scales, could be, for example, - is this person heavier or taller compared to other people? Of course, there would be no point in comparing kilograms (or pounds) to meters (or feet). However, we could compare the relative positions of that person on distributions of weight and height and determine whether his/her position relative to other people in the sample is higher on height than on weight. That is actually the way we make many real-life inferences – for example, for a person whose position on respective distributions is low on height, but high on weight, we would conclude that he/she is obese, while for a person average on weight, but high on height, we would conclude that he/she is slim and so on. To make these inferences using statistics, we could use percentiles or percentile ranks, and these statistics are indeed often used to make inferences like this. However, the problem with percentiles and percentile ranks is that they are on the ordinal level of measurement, so many further calculations with them, for example, those requiring addition or subtraction would mostly be impossible (or would not be meaningful, numbers tolerate anything, but the result loses meaning if things are not done right). A degree of improvement over percentiles for these purposes is afforded by the use of **standard scales and standard scores. Standard scales are scales with predefined properties**, typically with a **predefined mean and standard deviation and often also the shape of the distribution. Raw variable values** (values obtained through measurement or assessment) are then **transformed into standard scale scores.** The **process of transforming raw variable values (or raw scores) into standard scale scores is called standardization.** Because the properties of standard scales are predefined, **it is always known which standard scale score (or just standard score) corresponds to which position on a distribution i.e. to which percentile.** However, unlike with percentiles, standard scores are on the interval level of measurement, allowing for many calculations that could not be meaningfully performed on percentiles and percentile ranks (because percentiles and percentile ranks are ordinal measures!). Standard scores are widely used in professional practice. For example, the widely known IQ scores (used in the assessment of intelligence and other cognitive abilities) are a type of a standard scale score, with a predefined arithmetic mean of 100 and a standard deviation of 15 or 16 (e.g. Hedrih, 2020) and there are various others.

In statistics, **"standard scale" typically refers to the z scale** and standardization typically refers to the transformation of raw values into z scores (or to the z scale). **The z scale is a standard scale with an arithmetic mean of 0 and a standard deviation of 1. Z scores are also theoretically expected to be normally distributed i.e. calculations of theoretical z scores usually implies that they form a normal distribution** (although we may in practice encounter z scores obtained from empirical data forming a nonnormal empirical distribution!). **Data are converted to the z scale by subtracting the mean of the sample from each value and then dividing the result with the standard deviation.** The result is a z score:

z score = (Raw variable value–sample mean)/Standard deviation

Or **to convert z scores back to raw variable values we just multiply the z score with the standard deviation and add the mean:**

Raw variable value = z score * Standard deviation + Mean

Compared to raw variable values and percentiles, as indicators of the position on a distribution, z scores have multiple advantages:

- **Z scores are on the interval level of measurement** (unlike percentiles).
- **Z scores have no measurement unit,** i.e. the measurement unit is lost during the transformation when division is done with the standard deviation (standard deviation is in the same measurement units as the raw score), so the measurement units in the variable value and the standard deviation cancel each other. Because it has no measurement unit, **we can** than **meaningfully compare z scores on different variables,** variables **whose raw values are in different measurement units** (e.g. one in kilograms, the other in meters).
- **A z score indicates the magnitude of difference from the sample mean, expressed in standard deviations. One z score is one standard deviation.** A z score indicates how many standard deviations the value of the entity is above or below the sample mean.
- If the distribution is normal (or if it is known), we can precisely determine the percentage of entities below any z score i.e. **we can convert any z score into a percentile on the normal (or another theoretical) distribution.** Therefore, z scores are like percentiles, but on the interval level of measurement.
- It is easy to read – **positive values indicate values above the mean, negative values indicate positions below the mean.** Mean is 0. Unlike with raw variable values, it is not necessary to look up what the sample mean is in order to know whether a certain value is above or below mean and how far away from it.

Z scores are also components of various more complex statistics (some of the most basic of these calculations are presented in the later part of this book). They are actually one of the key components of many complex statistical procedures. It is also important to note **that the interval between z scores −1.96 and +1.96 contains 95% of entities of a normally distributed sample,** while **the interval between −2.56 and +2.56 contains 99% of entities in such a sample** (Figure 4.12).

4.8 Ipsatization

The standardization process described earlier refers to situations where values of all entities on a variable are converted to z score. When such an operation is done, it is said that that variable was converted to the z scale or that it was standardized. However, it is also possible to conduct the standardization procedure on values of a single entity on all measured variables or on a group of variables. Of course, for such a procedure to be meaningful, values of these different variables need to be comparable. This procedure is called **ipsatization.** In other words, **ipsatization is a procedure where values of an individual entity across a number of (comparable) variables are standardized i.e. converted to z scale. After ipsatization** is performed, **that particular entity,** whose values were ipsatized, **will have an average value of 0 on the group of variables that were**

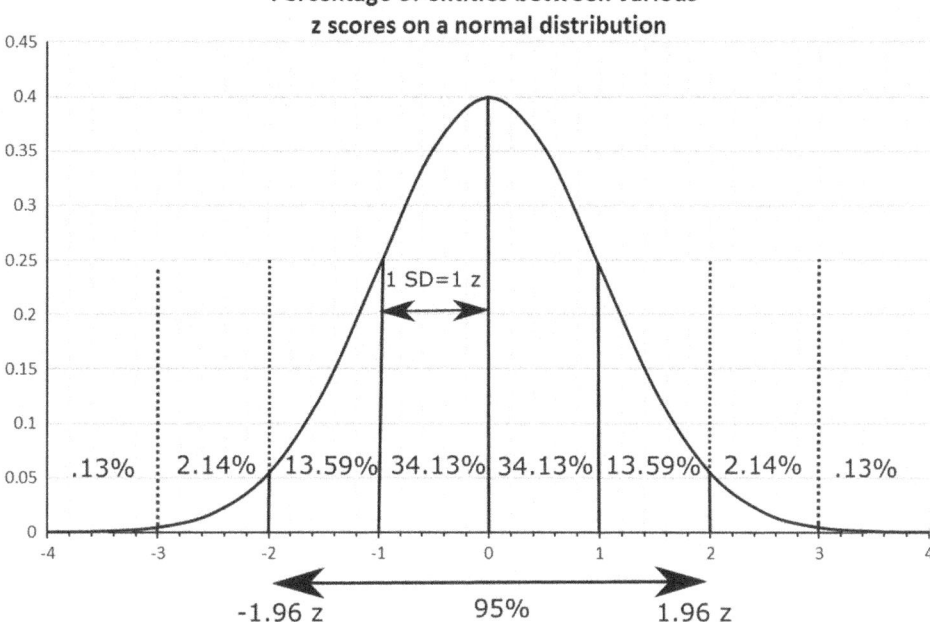

Figure 4.12 Percentage of entities on a normal distribution between various z scores. Units on the horizontal axis are z scores i.e. standard deviations. Numbers on the vertical axis are probabilities. We can see from the picture that, on a normal distribution, 34.13% entities lie between the mean (z score 0) and –1 z (the point that is 1 standard deviation below the mean). The same percentage of entities lies between mean and +1 z (the point that is 1 standard deviation above the mean). Also, 13.59% of entities lie between z scores 1 and 2 and the same percentage is between z scores –1 and –2. It is important to note that 95% of entities lies between z scores –1.96 and 1.96. The numbers for the 99% interval (not depicted) are –2.56 and 2.56 z. We can get the other percentages by adding and subtracting percentages between these intervals. For example, between z scores –1 and 1, there lie 34.13%+34.13=68.26% of all entities i.e. somewhat above two thirds of the whole sample (or population, depending on what is it we are considering).

included in the ipsatization procedure, and the standard deviation of values on these variables of 1. The ipsatization procedure is identical to the standardization procedure with the only difference being that standardization is done on values of all entities on a single variable, while ipsatization is done on values of all variables on a single entity.

When can ipsatizations be used in practice? Sometimes, it might not be important what value exactly an entity has on a particular variable, but what that value is like compared to values of that entity on other comparable variables. For example, when attitudes toward different topics are studied, **ipsatization can sometimes be used to help neutralize the effects of the response styles of study participants**. Imagine that you asked a group of people to rate how much they like various airline companies. Let us imagine that these people needed to rate these various airlines on a scale from 1 to 5 to indicate how much they like each one. Let us now imagine that some of these people really enjoyed their air travel experiences with these companies. They will then tend to give all these airlines 4s and 5s i.e. they will rate them highly. Let us then imagine that there are also people in that sample who generally hate flying or generally hated their experiences with these companies or who are simply much more demanding people. They will likely give all of them 1s and 2s. Now ask yourself this – does a 4 from someone who gave 5s to all other airlines really indicate a better attitude towards that particular airline than a 2 from someone who gave 1s to all the other airlines? And we can make this question even more practical – when he/she needs to travel by air, who is more likely to choose that particular airline – the one who gave it a 2, while giving all the other 1s, or the one who gave it a 4, while giving all the others 5s? The answer is most likely that the first one will choose it, in spite of the fact that he/she rated it lower (gave it a 2) than the second one (who gave it a 4).

Table 4.1 An example of 12 study participants who rated the service quality of 5 different airlines (fictional data). We can notice that, although the original answers of study participants are different, after ipsatization, five of them (marked in bold) have the same value. This is because the process of ipsatization converts raw scores into z scores that indicate the relative position of participant's answers in relation to all their other answers. Since all of them gave equal ratings to the 4 airline companies, while rating one company lower than the others, this resulted in equal positions on the distribution of answers regardless of specific ratings they chose. We should also notice that the one participant who gave 5s to all airlines, now has 0s on all of them, as his answers are a constant and thus all equal the mean of all his answers. We should also notice that although mean airline ratings given by participants are different, after ipsatization these are all equalized to 0. This is because ipsatization standardizes values of a participant on different variables, causing the mean of every participant across different variables that were included in the ipsarization to be 0 (while making the standard deviation 1)

How study participants rated airline companies, raw data Ratings were on a scale from 1 to 5, with 1 being the worst rating and 5 being the best

Study participant name	Airline A	Airline B	Airline C	Airline D	Airline E	Mean
Becky	3	5	4	3	2	3.4
Anita	5	5	4	5	5	4.8
Vladislava	4	2	5	5	2	3.6
Careen	2	2	1	2	2	1.8
Esmaeel	5	5	5	5	5	5
John	1	5	2	5	4	3.4
Dereck	5	4	4	5	4	4.4
Vladimir	3	1	3	5	4	3.2
Emmet	1	5	1	5	1	2.6
Mark	5	5	1	5	5	4.2
Ellen	5	5	3	5	5	4.6
Peter	4	4	2	4	4	3.6

	How study participants rated airline companies, ipsatized data					
Study participant name	Airline A	Airline B	Airline C	Airline D	Airline E	Mean
Becky	−.35	1.40	.53	−.35	−1.23	0.0
Anita	**.45**	**.45**	**−1.79**	**.45**	**.45**	**0.0**
Vladislava	.26	−1.06	.92	.92	−1.06	0.0
Careen	**.45**	**.45**	**−1.79**	**.45**	**.45**	**0.0**
Esmaeel	*.00*	*.00*	*.00*	*.00*	*.00*	*0.0*
John	−1.32	.88	−.77	.88	.33	0.0
Dereck	1.10	−.73	−.73	1.10	−.73	0.0
Vladimir	−.13	−1.48	−.13	1.21	.54	0.0
Emmet	−.73	1.10	−.73	1.10	−.73	0.0
Mark	**.45**	**.45**	**−1.79**	**.45**	**.45**	**0.0**
Ellen	**.45**	**.45**	**−1.79**	**.45**	**.45**	**0.0**
Peter	**.45**	**.45**	**−1.79**	**.45**	**.45**	**0.0**

Apart from this, there are various psychological tests that include the principle of ipsatization in their score calculation procedures – the sum of scores on all the test variables being fixed, with individual variables being able to vary with that restriction (e.g. Plutchik, 1989; Plutchik & Kellerman, 1974). Such tests are referred to as **ipsative tests.** There are also other examples, but nonetheless, it should be noted that, while standardization is a routine procedure, used practically everywhere, application of ipsatization is much less often encountered in practice.

4.9 Let us apply what we learned so far!

Let us now try to apply what we have covered in this chapter through a couple of exercises. Please refer to the start of the book for the general instruction for completing the exercises. Our suggestion is that you first read the excerpt and the statements and provide your own answer. You may write it in the Answer column, and after that look up the answers and compare your own answers with them.

Exercise F. Theoretical and empirical distributions, deviations from the normal distribution, standardization and ipsatization

Ellen is working as a consultant for a fruit processing company. This company is currently buying and equipping a cold storage building in which sorting, packing and storage of raspberries would be carried out. Ellen has to decide on which of the two raspberry sorting lines on offer the company should buy.

The first line presents each worker assigned to sorting with one raspberry at a time, and the worker should quickly inspect it and remove it from the line if it is bad. A worker of average vigilance working on this line has a 70% chance to spot a bad raspberry when it is presented to him/her (i.e. when the line presents a bad raspberry to a worker, the worker has a 70% chance to spot that it is bad). In one hour, this line exposes 5000 raspberries to each worker of which 1000 are typically bad.

The second line uses a wide conveyor belt that slowly moves and on which raspberries are spread out. Workers stand by the conveyor belt and rummage through passing raspberries trying to spot those that are bad. A worker of average vigilance will, on average, spot 15 bad raspberries per minute. In one hour, 5000 raspberries pass by each worker on average, out of which 1000 are bad.

The presented data are valid for workers of average vigilance. However, Ellen determined that there are pronounced individual differences between workers with regard to their vigilance.

It should be considered that values of all variables mentioned in the text or that can be derived from these data have distributions equal to ideal theoretical distributions expected for such variables and situations in which they are obtained. There are 60 one-minute intervals in one hour. With each line, each raspberry is inspected by only one worker (there are no situations in which the same raspberry passes by or is presented to more than one worker).

F	Statement:	Answer
F1.	The vigilance of female employees has the shape of Hong's distribution.	
F2.	If all workers had average vigilance, the distribution of the number of spotted bad raspberries per worker on the first line would have the shape of a Poisson distribution.	
F3.	The first line is more efficient in eliminating bad raspberries (i.e. it eliminates a higher percentage of bad raspberries).	
F4.	The second line processes more raspberries per hour than the first one.	
F5.	The vigilance of workers is uniformly distributed (has the shape of the uniform distribution).	
F6.	If all workers had the average vigilance, the average number of spotted bad raspberries per hour per worker on the first line would be greater than 650.	
F7.	If all workers had the average vigilance, the average number of spotted bad raspberries per hour per worker on the second line would be greater than 650.	
F8.	If we followed the work of a group of workers of average vigilance for two hours, working on the second line, the distribution of the number of spotted bad raspberries per worker would have the shape of a uniform distribution.	
F9.	The distribution of the number of spotted bad raspberries per worker on the first line, after only one raspberry has been exposed to each of them, would have the shape of the Bernoulli distribution (you should assume that all the workers have the average vigilance).	
F10.	Younger workers are more vigilant on average than older workers.	

Exercise G. (Popov et al., 2021) Theoretical and empirical distributions, deviations from the normal distribution, standardization and ipsatization

Table 1

Descriptive statistics for variables in the study

	Theoretical range	Achieved range	M	SD	Skewness	Kurtosis
Prior physical activity (GLTEQ)	0–119	0–119	38.83	28.88	.97	.48
Avoidant coping (Brief COPE)	12–48	12–43	21.40	4.90	.74	1.03
Problem focused coping (Brief COPE)	12–48	12–48	33.17	7.45	-.36	-.18
Emotion focused coping (Brief COPE)	6–24	6–24	15.48	4.24	-.22	-.02
Current physical exercise	1–4	1–4	2.57	.97	-.41	-.89
Depression (DASS–21)	0–21	0–21	4.07	4.87	1.50	1.75
Anxiety (DASS–21)	0–21	0–21	3.11	4.49	1.81	2.81
Stress (DASS–21)	0–21	0–21	7.38	5.64	.53	-.59

Table reprinted from: Popov, S., Sokić, J., & Stupar, D. (2021). Activity Matters: Physical Exercise and Stress Coping during COVID-19 State of Emergency. Psihologija, 54(3), 307–322. https://doi.org/10.2298/psi200804002p. Reprinted with the permission of authors.

G	Statement:	Answer
G1.	The distribution of Depression is positively asymmetric.	
G2.	The distribution of Anxiety is negatively asymmetric.	
G3.	The distribution of Anxiety is leptokurtic.	
G4.	The mean of Prior physical activity is lower than its median.	
G5.	The mean of Current physical exercise is lower than its median.	
G6.	The distribution of Anxiety is pointy, i.e. with study participants being more concentrated than is typical for the normal distribution.	
G7.	The distribution of Stress is platykurtic.	
G8.	The participants are more spaced apart on the Current physical exercise than would be expected if the distribution were normal.	
G9.	The 50th percentile of the variable Avodiant coping is higher than 21.40.	
G10.	The 80th percentile of Stress is higher than 22.	

Exercise H. (Hedrih, 2011; Holland, 1959) Theoretical and empirical distributions, deviations from the normal distribution, standardization and ipsatization

N=360	R	I	A	S	E	C
Skewness	1.38	−.44	.225	−.01	.30	1.51
Kurtosis	1.88	−.29	−.91	−.80	−.77	3.01
25th percentile	.18	.61	.69	.96	.69	.36
50th percentile	.44	1.05	1.21	1.45	1.19	.58
75th percentile	.80	1.50	1.82	1.99	1.76	.87

Prepared based on a part of the data used in Hedrih (2011).

R, I, A, S, E and C are measures of vocational interest types proposed by Holland's theory of vocational interests (Holland, 1959)

H	Statement:	Answer
H1.	If a person from this sample had a score of 1.5 on the variable I, his/her z score would be positive.	
H2.	There are no people in the sample who had a score higher than 2 on the variable E.	
H3.	There are more than 400 participants in this sample.	
H4.	The distribution of the variable S is symmetrical and platykurtic.	
H5.	The distribution of the variable C is positively asymmetrical and platykurtic.	
H6.	The distribution of the variable R has the shape of the Hansen's distribution.	
H7.	Median kurtosis scores are higher on the 25th than on the 75th percentile.	
H8.	Mean of the variable S is lower than 1.6.	
H9.	If a person from this sample had a score of 1 on S, his/her z score on this variable would be positive.	
H10.	The lowest score on variable A is 1.	

Let us now consider the answers:

F1 – meaningless. There is no such thing as a Hong's distribution.

F2 – false. No, the functioning of the first line is described in the terms of probability of an event (spotting a bad raspberry) per trial (exposure of the raspberry) and this is a setup for the binomial distribution. Hence, binomial distribution would be the expected distribution in this case.

F3 – false. Both lines process 5000 raspberries per hour/per worker and 1000 raspberries out of 5000 are bad. With the first line, a worker has a 70% chance to spot a bad raspberry when it is presented. It means that out of 1000 bad raspberries that will be presented to the workers they will on average spot 70%, which is 700. On the other hand, with the other line a worker will be spotting 15 bad raspberries per minute. There are 60 minutes in one hour. That means that a worker will, on average, spot 15 X 60 bad raspberries per hour, which equals 900. 900 hundred is higher than 700, which means that the second line is more efficient, eliminating 90% of bad raspberries, compared to the first production line that eliminates only 70%.

F4 – false. As noted in F3, they both process the same quantity per hour.

F5 – false. Vigilance (of workers) is a trait of individual differences and hence the expected distribution is normal, not uniform.

F6 – true. In f3, we already explained the calculation that it is 700 on average. 700 is higher than 650, hence the statement is true.

F7 – true. The working of the second line, as explained in f3, results in 900 spotted bad raspberries per hour per worker. 900 is more than 650, hence the statement is true.

F8 – false. The workings of the second line are explained in the terms of an average number of spotted bad raspberries per minute of work and that is a setup for the Poisson distribution, not a uniform one.

F9 – true. The workings of the first line are the setup for the binomial distribution. Bernoulli distribution is a special case of the binomial distribution when the number of trials is 1. Therefore. this statement is true.

F10 – unknown. Although this is certainly possible, there is nothing in the text to indicate whether this is true or not. Therefore, we do not know.

G1 – true. The skewness of Depression is 1.5, therefore positive. A positive skewness indicates that the distribution is positively asymmetric.

G2 – false. The skewness of Anxiety is 1.81, therefore positive. A positive skewness indicates that the distribution is positively and not negatively asymmetric.

G3 – true. Leptokurtic distributions have positive kurtosis. For anxiety, it is 2.81, therefore positive, meaning that the distribution indeed is leptokurtic.

G4 – false. We can see that Prior physical activity has positive skewness, which mean that its mean is higher than its median.

G5 – true. We can see that Current physical exercise has negative skewness, which means that its mean is lower than its median.

G6 – likely true. This statement is rather tricky. On the one hand, we can see that Anxiety has a pretty high kurtosis value, indicating that it is leptokurtic. On the other hand, we know that high kurtosis may also be a consequence of outliers and not of the extreme concentration of entities in the middle. However, if we compare its Achieved range (range of scores in the sample) we can see that the mean is less than 1 standard deviation above the lower boundary of that range, which means that a large part of the sample is concentrated in the rather small interval between the lower range boundary and the mean. We can conclude from this that there likely is an extreme concentration of participants in that low score area, although there is also a thick shoulder and tail of the distribution going to the positive side. Hence, the statement is likely true.

G7 − true. Kurtosis of stress is −.59, i.e. negative, meaning that the distribution is platykurtic.

G8 − true. "More spaced apart" implies a platykurtic distribution and this means that the kurtosis is negative. Kurtosis of −.89 indeed is negative, hence the statement is true.

G9 − false. 50th percentile is the median, skewness of this variable is positive, meaning that median is lower than the mean. The mean is 21.40 and the median must be lower than that, implying that the statement is false.

G10 − false. Although we have no data on the 80th percentile specifically, we can see that upper boundaries of both theoretical and achieved range of Stress are lower than 21. This means than there are no values higher than 21 and thus that the 80th percentile cannot be higher than 22.

H1 − true. We can see that skewness of variable I is negative, implying that the mean is below the median. Median is the 50th percentile, which is 1.05 in this case, hence the mean must be below that. All z scores above the mean are positive, meaning that the score of 1.5 must also be corresponding to a positive z value.

H2 − unknown. We can see that the 75th percentile on this variable is 1.76, but we do not know how far above this the range of scores extend. There might be someone with a score of 2, there might not be, we cannot tell from the table.

H3 − false. It says N=360. N is a typical designation for the number of entities in the sample and it is used in that way here. Hence, there are 360 entities in the sample, which is less than 400.

H4 − true. Skewness of variable S is practically 0, therefore the distribution is symmetrical. A negative kurtosis indicates that the distribution is also platykurtic.

H5 − false. Positive skewness indicates that it indeed is positively asymmetrical, however the positive kurtosis value indicates that it is not platykurtic, but leptokurtic.

H6 − meaningless. There is no such thing as a "Hansen's distribution".

H7 − meaningless. There is no such thing as "median kurtosis scores" that could meaningfully be said to have some bearing on the 25th and the 75th percentile. The entire statement is meaningless.

H8 − true. The variable S has a symmetrical distribution − skewness is 0, meaning that its mean and median i.e. 50th percentile have the same value. The value of the median is 1.45, and this is lower than 1.6

H9 − false. − We can see that the variable S is symmetrical as it has 0 skewness, meaning that its mean has the same value as its 50th percentile i.e. median. Since the median is 1.45, that means that a value of 1 would be below the mean and thus correspond to a negative z score.

H10 − false. We can see that the 25th percentile is .69, which is lower than 1. Hence 1 cannot be the lowest score on this variable.

References

Darlington, R. B. (1970). Is kurtosis really "peakedness?" *American Statistician*, 24(2), 19–22. 10.1080/00031305.1970.10478885

de Moivre, A. (1756). *The Doctrine of Chances: Or, A Method of Calculating the Probabilities of Events and Play*. Chelsea Publishing Company.

Field, A. (2009). *Discovering Statistics Using SPSS*. SAGE Publications Ltd.

Hahs-Vaughn, D. L., & Lomax, R. G. (2020). *An Introduction to Statistical Concepts - 4th Edition*. Routledge.

Hair, J. J., Hutt, T. G., Ringle, C., & Sarstedt, M. (2016). *A Primer on Partial Least Squares Structural Equation Modeling PLS-SEM*. SAGE Publications, Inc. https://www.amazon.de/gp/product/B01 K0RVE0C/ref=as_li_tl?ie=UTF8&camp=1638&creative=6742&creativeASIN=B01K0RVE0C& linkCode=as2&tag=httpwwwsma079-21

Hamburger, A., Hancheva, C., & Volkan, V. (Eds.). (2021). *Social Trauma - An Interdisciplinary Textbook*. Springer Nature Switzerland AG. 10.1007/978-3-030-47817-9

Hedrih, V. (2011). Provera konvergentne i diskriminativne validnosti analizom multiosobinske-multimetodske matrice na primeru PGI testa profesionalnih interesovanja zadatom uzorku iz Republike Makedonije [Estimation of convergent and discriminant validation by the multitrai. *Primenjena Psihologija*, *4*, 393–408. http://primenjena.psihologija.ff.uns.ac.rs/index.php/pp/article/ view/1138/1152

Hedrih, V. (2020). *Adapting Psychological Tests and Measurement Instruments for Cross-Cultural Research: An Introduction (1st Edition)*. Routledge, Taylor&Francis Group.

Holland, J. L. (1959). A Theory of Vocational Choice. *Journal of Counseling Psychology*, *6*(1).

Plutchik, R. (1989). Measuring Emotions and their Derivatives. In *The Measurement of Emotions* (pp. 1–35). Elsevier. 10.1016/B978-0-12-558704-4.50007-4

Plutchik, R., & Kellerman, H. (1974). *Emotion Profile Index*. Western Psychological Services.

Popov, S., Sokić, J., & Stupar, D. (2021). Activity Matters: Physical Exercise and Stress Coping during COVID-19 State of Emergency. *Psihologija*, *54*(3), 307–322. 10.2298/psi200804002p

Tošković, O. (2020). *Autostoperski vodič kroz statistiku: Uvod u primenjenu psihologiju [Hitchhikers guide through statistics: Introduction to Applied Statistics]*. Centar za primenjenu psihologiju.

Westfall, P. H. (2014). Kurtosis as Peakedness, 1905–2014. R.I.P. *American Statistician*, *68*(3), 191–195. 10.1080/00031305.2014.917055

5 Inferential statistics, basic concepts

Inferential statistics refers to a set of methods used to make inferences about the properties of the population based on sample statistics i.e. to infer about the values of parameters based on values of statistics. In the previous parts of this book, we presented various procedures for describing a sample, both as a whole and of values of individual entities in comparison to the rest of the sample and also for collecting a sample. But, most of the time, the point of using statistical procedures is not to discover things about a specific sample, but to use that sample to make inferences about the population. The issue with this is that we know that a sample taken from a population, even using the best available sampling procedures is not guaranteed to be totally representative i.e. it is not guaranteed that all of the properties of the sample will be exactly like those of the population. On the contrary, it is quite likely that the properties of the sample might differ somewhat. This is why, when making inferences about the population, this possibility that the sample properties will be more or less different to those of the population has to be accounted for. That is one of the reasons, why we draw a distinction between statistical indicators calculated from the sample and those same indicators in the population, where the former are called statistics and the latter are called parameters.

At the moment this book is written, two approaches to addressing the issue of assessing values of parameters based on values of statistics are in most common use: **the approach based on the central limit theorem** and the **approach based on the use of resampling**, primarily on the use of **bootstrapping**.

5.1 The central limit theorem

The central limit theorem postulates what happens when we take multiple samples from the same population, which is actually a typical situation in which research findings are verified – one study takes a sample from the population to be studied and reports its findings and, later, another study takes another sample from the same population and reports whether their findings match the findings reported by the first study.

The central limit theorem postulates that if we take a large number of random samples from a population and we then calculate the same statistic from each of these samples, the distribution of values of this statistic across these samples will approximate the normal distribution. When postulating this, it is also assumed that either the population in unlimited or very large compared to the sample (to be practically the same as unlimited) or that the sampling is done with replacement (so that probabilities of individual entities for being included in the sample remain constant throughout the sampling process). For example, this means that we

DOI: 10.4324/9781003107712-5

could take, for example, 2000 random samples from the same population, measure a certain variable on each sample and then calculate a certain statistic from each of those samples. For example, that statistic could be the arithmetic mean on the variable that was measured. We would therefore calculate the arithmetic mean on that variable of each of these 2000 random samples. We would than examine the distribution created by these 2000 means (1 mean from each sample equals 2000 means!) and we would discover, according to the central limit theorem, that this distribution has the shape of a normal distribution.

This **distribution of values of statistics calculated from this large number of random samples is called the sampling distribution. According to the central limit theorem, the arithmetic mean of the sampling distribution is equal to the parameter** (Figure 5.1).

The postulate that the sampling distribution mean is the parameter implies that, **although statistics of individual samples will be more or less different from the parameter, they will still tend to cluster around the parameter.** And the parameter is also the center of the sampling distribution, given that it is the mean and that the sampling distribution is normal and, thus, symmetrical. The more a sample statistic differs from the parameter, the less likely it is that such a sample could be obtained through random sampling. **The location of the parameter on the sampling distribution is also the point of highest probability** i.e. if we needed to guess where on the sampling distribution the statistic of a random sample we drew from the population is, we would make the least mistakes if we assumed that it is located at the center i.e. that it is the same as the parameter. By making that guess we would, indeed, more often be wrong than

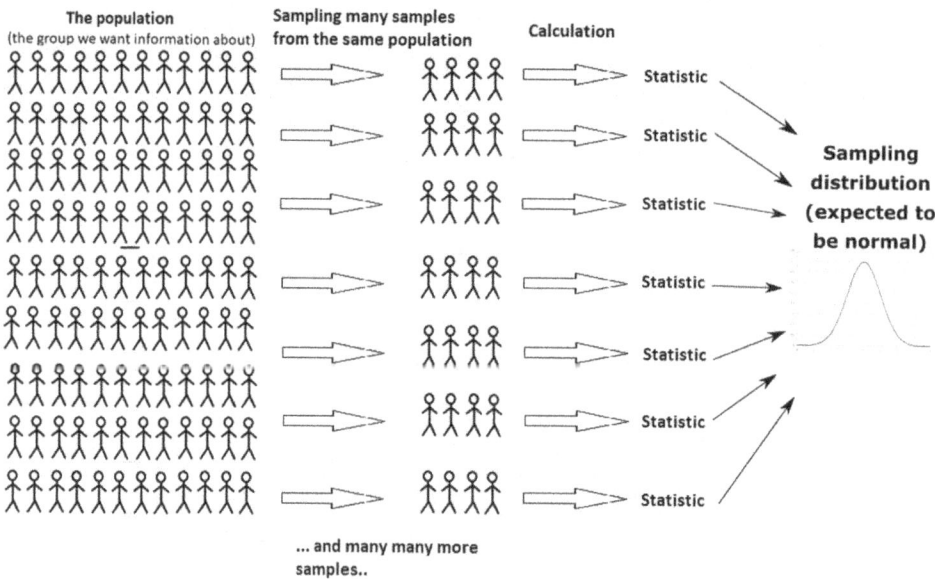

Figure 5.1 A graphical representation of the key postulates of the central limit theorem. A large number of samples is taken from the same population and the same statistic is calculated from each of them. The distribution of all the statistics obtained in this way is called the sampling distribution and the theory postulates that it will have the shape of the normal distribution and that its mean will be equal to the parameter. The standard deviation of the sampling distribution obtained in this way is called the standard error.

correct, but we would be more often correct than if we assumed it to be at any other point, because the center of the distribution is the point of highest probability i.e. the point with the greatest concentration of entities. Also, it is the point that has the smallest average distance from any other place in the distribution (because it is at the center), so **the average size of our miss would still be smallest if we assumed that a sample statistic, whose position in the sampling distribution is unknown, is in the center of the distribution** and thus equal to the parameter.

Aside from the mean i.e. a measure of central tendency, a description of a distribution requires a measure of variability or in this case – the standard deviation. **The standard deviation of the sampling distribution is called the standard error.** It basically tells us how different can statistics of individual samples be from population parameters. Also, if we refer to the chapter on standardization, we will see that there is a clear relationship between deviation from the mean expressed in standard deviations and the position on the distribution. This means that **the size of the standard error can tell us how probable are various difference sizes between the sample and the population.** However, to use the standard error in this way, we need to calculate it first. We know that for calculating a standard deviation we need individual values of entities in the sample. In this case, that would be the multitude of samples taken from the same population as postulated by the central limit theorem and this is something that is not actually available in practice – what we typically have is the one sample we collected for the purposes of the ongoing study. Due to this, **under the central limit theorem approach, standard errors cannot really be calculated from data** (because we have only the one sample collected for the particular study in the scope of which we are doing the calculations, not the required multitude of samples!), **but are estimated based on formulas. There is a specific formula for calculating the standard error of every statistical indicator,** but **common to all of them is that the size of the standard error is dependent on the standard deviation of the sample on the examined variable and the sample size. The higher the standard deviation, the higher the standard error. The larger the sample the smaller the standard error.** The more variability there is in the sample (and hence likely in the population) the more opportunity for the sample to be different from the mean. For an obvious example, if everyone in a population was exactly 2 meters tall, there would be no chance to select a sample from such a population that would be different. On the other hand, if the heights of people in the population varied, it would indeed be possible to select a sample that has more short than tall people or vice versa. Likewise, if we randomly select people from a population, but are, for example, selecting just one person, there is a certain probability that we, by random chance, select a person that is very tall and thus not at all representative of the typical height of the population. However, if we randomly select a multitude of people from the population, the chance that all of them will happen to be very, very tall and thus very, very different and un-representative of the typical height of the population is much, much smaller (although never zero for unlimited populations and/or sampling with replacement!). It is the same situation as with winning jackpots on lotteries such as Lotto for example. While it is not too difficult to find a person who won a Lotto jackpot (which is a very rare event, but every now and then someone wins it), there is likely no person that won 100 jackpots in a row on the same lottery (an event with probability so low that it will likely never happen). This is why increasing sample size decreases the standard error.

Formulas for calculating standard errors of various statistics are generally derived from the postulates of the central limit theorem and they differ for different statistics. For example, the formula for the **standard error of mean is** obtained by dividing the standard deviation by the square root of the number of entities in the sample (e.g. Harding et al., 2014):

$$SE_{mean} = \frac{SD}{\sqrt{N}}$$

Standard error is typically denoted by SE with the notation or the name of the statistic to which the standard error refers to in subscript. In this formula, SE_{mean} is the standard error of mean, SD is the standard deviation of the sample and N is the number of entities in the sample.

The formula for the **standard error of the median** (Harding et al., 2014) is:

$$SE_{median} = 1.253 * \frac{SD}{\sqrt{N}}$$

We can see from this that **the standard error of the median is approximately a quarter larger thatn the standard error of the mean** given the same standard deviation and sample size. The **standard error of the standard deviation** is (Harding et al., 2014):

$$SE_{SD} = \frac{SD}{\sqrt{2(N-1)}}$$

We can see from the formula that, with the same mean and sample size, the standard error of the standard deviation is lower than the standard error of the mean. Given that $1/\sqrt{2}$ equals roughly .71 and that the rest of the formula is practically the same as the formula for the standard error of mean, we can conclude that **the size of the standard error of standard deviation is a bit above 70% of the size of the standard error of mean**.

The formulas for **standard errors of skewness and kurtosis** are somewhat different from the formulas presented so far in that they **are estimated solely based on the sample size and their calculation does not require the standard deviation of the sample**. That means that these values will be equal for all samples of the same size. **The standard error of skewness** can be estimated as (Harding et al., 2014):

$$SE_{skewness} = \sqrt{\frac{6N(N-1)}{(N-2)(N+1)(N+3)}}$$

where N is the number of entities in the sample i.e. the sample size. The **standard error of kurtosis** is (Harding et al., 2014):

$$SE_{Kurtosis} = 2 * SE_{Skewness} * \sqrt{\frac{N^2 - 1}{(N-3)(N+5)}}$$

Another important statistic that it is important to mention here is the correlation coefficient (Pearson correlation coefficient, presented in the later part of this book). The formula for estimating **the standard error of the Pearson correlation coefficient** (typically denoted by r) also does not include the standard deviation, but it does include the correlation coefficient itself and it looks like this:

$$ SE_r = \frac{(\sqrt{(1 - r^2}}{(\sqrt{(N - 2)}} $$

In this formula, r is the size of the correlation coefficient. It should be noted **that in most common practical applications**, that are discussed in the later part of this book, the r coefficient in this formula is zero, so **the standard error of the correlation coefficient roughly** (if we neglect the -2 part of the formula) **equals the reciprocal of square root of the number of entities in the sample.**

During the past century and also at the time this book is written the central limit theorem is the most widely used approach in inferential statistics. It is applied in almost all tests and statistical procedures for making inferences about population values, for making inferences about differences between populations and the assumption of the normal sampling distribution is even used when evaluating results of multiple studies of a topic to draw conclusions whether there were any irregularities with reporting results of studies. The assumption is that when everything is regular, statistics from a multitude of studies of the same phenomenon would result in a normal distribution of statistics from different studies. Irregularities changing this situation are commonly considered to come in the form of the so-called "drawer effect". **"Drawer effect" is** a situation **when results of studies that are considered undesirable are not published** (either because researchers do not publish them or because publishers refuse to publish them), **while the desirable ones are published**. Therefore, statistical procedures for analyzing the shape of the distribution of results of a multitude of studies to identify whether some parts of the sampling distribution are missing are applied and the deviation of the shape of the distribution of statistics of multiple studies from the normal distribution are taken as an indicator of possible irregularities. Also, practically all mainstream statistical software packages in use at the time this book is written apply primarily inferential procedures that rely on the assumptions of the central limit theorem.

5.2 Bootstrapping approach to inferring parameter values

In spite of the widespread use of procedures based on the central limit theorem, an important issue with it is that it assumes that the sampling distribution will be normal, without actually verifying that it indeed is in every particular case when such assumptions are applied. In other words, the problem is that the central limit theorem is rarely empirically tested, yet is widely applied. Therefore, alternative methods for calculating standard error and making inferences about the population have been proposed. An overview of some of these methods can be found for example in Efron (1981).

At the moment this book is written, the most traction in the scientific community and with the makers of statistics software seems to have obtained the **bootstrap procedure** for making inferences about the population. **An advantage of the bootstrap procedure over the central limit theorem is that it does not make any assumptions about the shape of the sampling distribution.** Instead, the bootstrap procedure, as typically used for these purposes, solves the problem of the missing sampling distribution, by taking a multitude of samples from the study sample instead of from the population. **In the bootstrap procedure, a large number of samples** (e.g. 10000 samples, the number is limited only by the processing power of the computer used for this and the willingness of the researcher to wait for the results) **is taken with replacement from the study sample**. In this way, because sampling is done with replacement, it is possible to take an unlimited number of samples from the same, limited study sample. **The study sample in this procedure serves as a proxy for the population, while the large number of samples taken from the study sample using the bootstrap method are taken to be a proxy for the large number of samples taken from the population**. The same statistic is then calculated from each of these samples created through bootstrapping and **the distribution of values of that statistic across the bootstrap samples is taken to be the sampling distribution. The standard deviation of the sampling distribution thus created is considered the standard error. The statistic of the study sample** (the actual sample from which all the bootstrap samples were created) is **typically expected to be around the center of the bootstrap distribution**, but **this is sometimes not the case**. When doing inferential procedures based on bootstrapping, **the difference between the mean of the sampling distribution created via bootstrapping and the mean of the original study sample is typically reported**. It is most often referred to as the **bias,** but other names such as **deviation or simply difference** can also be encountered.

So, unlike the situation when applying the central limit theorem, where we just estimate the properties of the sampling distribution (primarily the standard error) using formulas, in the bootstrapping approach the sampling distribution is created in the way described above and the standard error is calculated directly.

5.3 Assessing population parameters, point and interval assessment

There are two primary approaches to estimating parameters – the point estimation approach and the interval estimation approach. These two approaches to estimating parameter values can be applied both in the scope of the central limit theorem and through the bootstrapping procedure. In the point estimation approach, an estimate of the parameter is reported, along with the standard error. In the interval estimation approach, the parameter is estimated by defining a confidence interval and specifying the probability that that confidence interval contains the parameter.

In **the point parameter estimation using the central limit theorem** the parameter is assumed to equal the statistic and the standard error of the statistic is reported. Of course, the postulates of the central limit theorem tell us that the statistic calculated from our study sample can be anywhere on the sampling distribution. However, not all positions on the sampling distributions are equally probable and the most probable one is directly at the center of the distribution i.e. at the place where the parameter is. That is the reason why the **point estimation in the scope of the central limit theorem approach is done by assuming that the statistic equals the parameter.** As a proper description of a sample includes specifying a measure of central tendency and a measure of variability, it is also necessary to provide one such measure, that would account for the likely size of the error of estimation (the error we make when we assume that our statistic equals the parameter). This is done by reporting the standard error. Therefore, **a proper point estimation of the parameter** in the scope of the central limit theorem **is done by assuming that our sample statistic equals the parameter and by reporting the standard error of the statistic.** As mentioned in the previous subchapter, in the approach based on the central limit theorem, standard errors are estimated using formulas.

The **point estimation based on the bootstrapping** approach is performed by reporting the difference between the statistic of the study sample and the mean of the sampling distribution created through the bootstrap procedure, called the bias and the standard deviation of this sampling distribution i.e. the standard error. In simpler words, **in the bootstrap approach, point estimation of the parameter is done by reporting the sample statistic, the mean of the bootstrap sampling distribution, the bias and the standard error.**

The **interval estimation of parameters** is done by defining confidence intervals that have a defined probability of containing the parameter. This is done by **defining and estimating an interval that would include a predefined proportion of cases from the sampling distribution**. In the central limit theorem approach, estimation of the confidence intervals is based on the expectation that the sampling distribution is normal (even though we do not actually have the sampling distribution!) and **confidence intervals are created by defining an interval whose upper and lower border are equally distant from the mean and which would contain a desired percentage of entities on a normal distribution** the mean of which is the sample mean and the standard deviation of which is the standard error. As explained in the previous chapter, differences from the mean expressed in standard deviations are z scores and z scores can always be converted to percentiles. This means that **we can define an interval on the normal distribution that contains a desired proportion of entities and the borders of which are equally distant from the mean, by simply defining the z scores** (i.e. number of standard deviations below or above the mean) **between which certain proportion of entities is located on the normal distribution.**

In the bootstrapping approach, the bootstrap procedure allows us to create the sampling distribution and then **we simply define the interval on the sampling distribution that we want to create confidence interval from.**

The **most commonly used proportions of the sampling distribution** to be used for the creation of the confidence intervals **are 95% and 99%.** Although it is equally possible to use any other percentage, there is a strong custom among researchers to apply one of these two percentages.

Under the central limit theorem, **to create the 95% interval we should multiply the standard error with 1.96 and for the 99% interval the number is 2.56,** because on the normal distribution 95% of entities are located between z scores −1.96 and +1.96 and 99% of scores are between z scores −2.56 and +2.56. **An interval of 1 z score equals one standard deviation, or one standard error for a sampling distribution,** so we could also say that 95% of entities on a normal sampling distribution are located in the interval that starts 1.96 standard errors below the mean and ends at the point that is 1.96 standard errors above the sample mean. So to establish a 95% confidence interval around a statistic, under the assumptions of the central limit theorem we would use the following formula

$$CI_{upper\ limit} = Statistic + 1.96\ SE_{statistic}$$
$$CI_{lower\ limit} = Statistic - 1.96\ SE_{statistic}$$

In this formula CI refers to upper or lower limit of the confidence interval and this is generally an abbreviation very often used in scientific literature to denote confidence intervals. The statistic could be any statistic that we are using to make inferences about the corresponding population parameter. That statistic could be the mean, median, standard deviation, variance, skewness, kurtosis or any other sample statistic. SE is the standard error of that statistic.

To calculate the 99% confidence interval under the central limit theorem, we would use the same formulas, but would only replace 1.96 with 2.56. It is also possible to calculate any other percentage interval desired and this is done in the same way with the only difference being that the coefficient that is 1.96 for 95% interval would have to be replaced with a corresponding coefficient for that desired interval. This coefficient can be calculated in statistical software using functions that convert desired positions on the normal distribution into z scores corresponding to them.

In the **bootstrap approach** to parameter estimation, the sampling distribution is available (actually, a proxy of the sampling distribution – the distribution of statistics from the large number of samples drawn from the study sample, as described above), **so a confidence interval is created by simply identifying an interval that contains the desired percentage of entities** i.e. statistics from the sampling distribution. This interval is created so that **its upper and lower boundary are equally distant from the center of the bootstrap sampling distribution** (Figure 5.2).

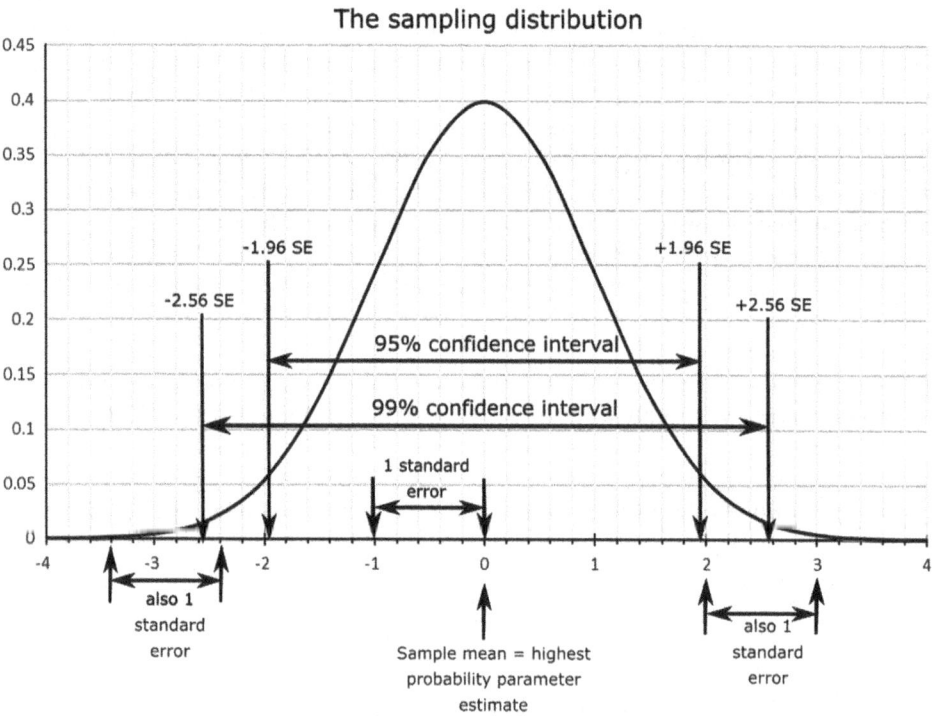

Figure 5.2 A graphical representation of the sampling distribution and related concepts. Positions of the 95% and 99% confidence intervals on the sampling distribution as well as the size of one standard error relative to the distribution and the confidence intervals are presented. Numbers on the vertical axis are probabilities, while the numbers on the horizontal axis are variable values in z scores. The size of a z score equals one standard deviation and, given that this is a sampling distribution, a standard deviation is called a standard error, so we can say that units on the horizontal axis are also standard errors.

What happens if we happen to be very wrong with the assumption that the statistic equals the parameter? This is especially important given that we will not know whether that is the case when making inferences based on our sample. What if our statistic is not very close to the parameter? Yes, it is indeed possible and will often be the case that the considered statistic of the study sample and the parameter we wish to estimate are not really close. The central limit theorem describes exactly such possibilities. It will also sometimes happen that the statistic of our study sample is somewhere on the outermost parts of the sampling distribution and as we do not really know the value of the parameter, based on that single study, we will not know about this. However, that is where the standard error and confidence intervals come into play and we should just remind ourselves of the obvious - the distance of the statistic from the parameter equals the distance of the parameter from the statistic. So, if we set up a confidence interval around the parameter (let us suppose that we knew the value of the parameter it; setting up around it means that we place the confidence interval so that the parameter is in its center) and such an interval was wide enough to include the statistic from our sample, than a confidence interval of equal width that is set around the statistic (with the statistic in its center) will be wide enough to include the parameter! Therefore, as long as our confidence interval is wide enough, everything should be OK most of the time and in case of point estimation the size of the standard error will give us an idea of the magnitude of the likely error in estimating the parameter and we should work onward with the understanding of the level of precision with which we are estimating the parameters. That said, it is also a fact that, **on rare occasions, our parameter estimates will be more substantially off** and that possibility is included in the methodology applied **– a 95% confidence interval functioning as expected still means that there will be 5% of cases when the parameter will not be within our confidence interval (or 1% for a 99% confidence interval).**

5.4 The null hypothesis

When we do research on a certain phenomenon, aside from describing it, or in the case of statistics, making inferences about properties of the population based on the studied sample or samples, we typically also want to be able to draw certain conclusions about the relationship between the phenomenon studied and other known or similar phenomena. For example, when we study certain properties of a certain population, we will not stop at establishing those properties based on our sample, but we can also ask ourselves whether the properties of the population we are studying are different from the properties of another known population. Are people from the country A taller than people from country B? Are attitudes of people in this city different from attitudes of people from a different city? Do children of different age have different levels of expression of a certain psychological trait? Or we can be curious about relationships between different phenomena. For example, we may want to find out whether a certain personality trait makes people more likely to make certain business decisions or does it make them more likely to buy and keep certain items. Or we may want to know whether a certain action produced effects. For example, we may ask ourselves whether the psychotherapeutic treatment we applied leads to changes in people who underwent that treatment. Or whether knowledge about a certain topic increased in a group of students that took a course about it? Or whether a medicine works i.e. reduces symptoms of a certain disease or whether it cures it completely?

All of these questions are typically answered by first specifying a hypothesis, which is, essentially, a specification of our expectation about the state of relations between phenomena we are interested in (variables, groups/populations) and then testing whether they hold. However, to test a hypothesis, we need that hypothesis to be very precise while, most of the time, our expectations about the state of relations between variables or groups of entities are not precise at all. For example, our expectation that people will know more about a certain topic after taking a study course, need not include a specification on how much exactly the knowledge level of the group that took the course will increase (in comparison to their level before the course or to some group that did not take the course). Our expectation that a medical procedure will work might not necessarily mean that 100% of people with a certain ailment undergoing the procedure will be healed. A procedure that heals 80% of people might still be considered effective in many cases. Similarly, we can conclude that attitudes about a certain topic of two populations differ both in the case when these attitudes are completely different and in a situation when they differ somewhat. So, specifying the precise degree of difference between groups to be compared or the strength of relationship between variables whose relations we are studying is often not feasible. For example, if our hypothesis states that applying a certain medical procedure will heal 100% of people with a certain ailment and after testing that hypothesis find out that 100% of people were not healed, does this mean that the procedure does not work at all or that it does work but not in 100% of cases? Or if we discover that taking a study course does not result in precisely, say, a 55-point average increase on a certain test of knowledge, does that mean that people did not learn anything on that course or that their change in the level of knowledge just corresponds to some other level of change of the score on the knowledge test. Might be both, of course. These are types of situations where the null hypothesis comes into play as a useful tool.

The way it is most often used and defined in social sciences research, a null hypothesis is any hypothesis that states that the value of the considered parameter (or of the difference between parameters, which is a parameter in itself) is 0. The null hypothesis is typically denoted as H_0. It can state that two samples come from populations whose means on the examined variable differ by 0 (i.e. that they do not differ). Or that the value of a certain statistic in the population is 0. Or that examined samples come from populations whose examined statistics differ by 0 (i.e. are all the same) etc. The value of 0 is an important value because if the difference between two values is 0, we state that they are the same, but if it is not, we can conclude that they differ. Also, if the size of a certain effect is 0, then there is no effect. If the effect is anything other than 0, this means that that the effect exists. **By falsifying the null hypothesis, we demonstrate that what is claimed by the hypothesis is not zero.** That there is a difference between populations, that the examined effect exists (is not 0), that parameters we were comparing differ etc. **The most common use of the null hypothesis is to examine whether two or more samples come from the same population (or from populations between which the difference** is 0 i.e. that are the same with regard to the examined property) **or that some statistical parameter is 0.**

However, **as originally conceptualized, H_0 or the null hypothesis need not claim that the value is zero, but may be any exact statistical hypothesis i.e. it may specify any other exact value or state of relations as long as it is precise. The null hypothesis is so called because it is conceptualized as a hypothesis that stands to be nullified with empirical data** (e.g. Gigerenzer, 2004), not from the word null meaning zero. Due to this, **some authors pull a distinction between a null hypothesis**, which they define as any exact statistical hypothesis that stands to be nullified by testing **and the nil hypotheses**, which they define as a subset of null hypotheses that specify that the considered parameter or relationship between parameters equals zero (essentially, what is typically referred to as the null hypothesis, these authors refer to as the nil hypothesis).

Why use the null hypothesis at all? As described in previous examples, **the main advantage of the null hypothesis is that it is precise – it states that something equals zero**. On the other hand, the so-called **alternative hypothesis,** the hypothesis describing our real expectations, **is typically not precise**. The alternative hypothesis might be an expectation that statistics differ, that the considered statistic is a non-zero value, that the studied effect exists or something similar, but a difference might mean many different levels of difference, a non-zero value might mean many different values and also an effect that exists might be of many different sizes, from a very low to an extremely high effect. **The null hypothesis and the alternative hypothesis are opposites to each other, so if one of them is true, the other is not.** This means that by testing the null hypothesis, which is precise and therefore testable, we are also testing the alternative hypothesis. If we test the null hypothesis stating that a treatment we are interested in has a 0 effect i.e. that it has no effect, and our results show that it is true, that means that our alternative hypothesis i.e. the expectation that the treatment has an effect is not true. On the other hand, if we conclude that our null hypothesis is not true, that implies that our alternative hypothesis, our real expectation is valid. In the previous example with testing for the effect of a treatment, concluding that the null hypothesis is not true i.e. rejecting it, would imply that our treatment has a non-zero effect.

How do we apply this in practice? **The general procedure for testing the null hypothesis starts by assuming that the null hypothesis is true in the population**. We remember from the previous chapter that sample statistics can deviate more or less from population parameters. So, this procedure starts by assuming that the null hypothesis holds in the population regardless of the values of statistics we obtained on our sample. Then, **we calculate the probability that the results we obtained on our sample or results deviating more from the null hypothesis be obtained in a situation when the null hypothesis is true in the population**. We make this calculation either by applying the central limit theorem or directly through resampling methods, as described in the previous chapter. Finally, **based on that probability** i.e. on how high that probability is, **we decide on whether we find it probable that the null hypothesis holds in the population or not.**

For example, if we wanted to test whether two samples come from the same population, or more precisely, from populations that have the same mean values on the variable we are interested in, we start by assuming that the difference between the parameters of the two populations i.e. means on that variable is 0. We than calculate the probability, either according to the central limit theorem or through resampling, that the difference between means of the two samples that was obtained on the sample or larger, be obtained on two samples when the null hypothesis holds in the population. In the end, we decide whether the probability we calculated is sufficiently small for us to conclude that it is not probable that the null hypothesis is true in the population, leading us to reject the null hypothesis and accept the alternative one or that this probability is sufficiently large that we declare that it is likely that the null hypothesis is true in the population thus leading to the acceptance of the null hypothesis and rejection of the alternative one.

A key statistic in these considerations is the probability to obtain the results that were obtained on the sample or more extreme (i.e. that deviate more from the null hypothesis than those obtained on the sample) **in a situation where the null hypothesis is true in the population. This statistic is called statistical significance. Statistical significance is typically presented as a proportion, with values between 0 and 1. It is usually designated as p. or sig. and often referred to as the "p value" or just "significance".**

For example, if we wanted to infer whether two samples come from populations with the same mean on the considered variable, we could calculate the probability that the difference between means of two samples as large as was obtained between our samples or larger be obtained in a situation when the difference between population means is zero. That probability would be statistical significance. The same would be the case if we calculated the probability that a certain sample statistic (for example, skewness or kurtosis) has the value it has on our sample or larger (actually – more different from zero, either in the positive or the negative direction) in the situation when the population value is zero.

Statistical significance is used to decide whether to reject or accept the null hypothesis. This is typically done by comparing the value of statistical significance with the critical level we have accepted. **The most typically used critical levels of statistical significance found in literature are 5% and 1% or .05 and .01. Applying the 5% or .05 critical significance level means** that if the probability of **obtaining the results such as those obtained on the sample or more extreme** in a situation when the null hypothesis is true in the population **is 5% or below 5% (p<.05),** we will then **reject the null hypothesis**. On the other hand, if the probability of **obtaining the results such as those obtained on the sample or more extreme** in a situation when the null hypothesis is true in the population **is greater than 5% (p>.05)** than **we accept the null hypothesis.** "More extreme results" here means results that differ from the situation specified by the null hypothesis more than the results obtained on the sample. **The procedure for deciding whether to accept or reject the null hypothesis using the statistical significance level of .01 or 1% as the threshold is the same, only the critical level is now the statistical significance of .01 instead of .05.**

Statistical significance is generally calculated in the same way confidence intervals are calculated when assessing parameters (please see previous chapter) – we create the sampling distribution that would be obtained if the null hypothesis were true or calculate its parameters and then we designate the area of the distribution containing values of statistics that were obtained on the sample and those that differ more than that from the values specified by the null hypothesis (i.e. 0). The proportion of cases on the sampling distribution falling into that equals statistical significance. For example, if we wanted to compare means of two populations and our null hypothesis was that the difference between population means is zero, we would than create or calculate the properties of the sampling distribution of differences between means of samples taken from these two populations. If, for example, the difference between means of samples we are comparing is, say 2, we would than calculate the proportion of cases on this sampling distribution of mean differences that equal two or above and this proportion would represent the statistical significance level of this difference.

That said, there are generally two ways to designate this area when calculating statistical significance – we can calculate **a one-tailed significance level or a two-tailed significance level.** One-tailed significance is calculated by including into the area for assessing statistical significance only cases that are on the same side of the sampling distribution as the results expected on the sample, while the two-tailed significance is calculated by including cases with the same or higher level of difference from the value specified by the normal distribution on both sides of the distribution. In the previous example, where the null hypothesis specifies the difference between means of two populations to be zero and the difference between means of the two samples from these two populations turned out to be 2, a one-tailed statistical significance would be obtained by marking only the area of the sampling distribution containing values of 2 and above. A two-tailed sampling distribution would be obtained by including values of 2 and above (one side of the distribution – the side above the mean) but also values of −2 and below (the other side of the distribution – the side below the mean). This is because values of 2 and −2 represent here the same level of difference from the situation specified by the null hypothesis, only in different directions – one in the positive and the other one in the negative direction. This means that, given a symmetrical sampling distribution, the two-tailed statistical significance level will always be twice the size of the one-tailed significance level – if the one-tailed significance level is .1 the two-tailed significance level will be .2 (Figures 5.3 and 5.4).

When do we use one-tailed and when two-tailed distribution? **In practice, researchers will always calculate two-tailed significance levels except in cases when deviations from the situation defined by the null hypothesis are possible only in one direction, when differences from the values specified by the null hypothesis in the other direction are impossible.** This can be the case only when there are clear theoretical reasons why differences in the other direction are impossible or in situations where it is not technically possible for the values to deviate from the null hypothesis in one of the directions. For example, if we had a course on a certain topic in a study program, a topic students knew nothing about prior to the course, we could meaningfully calculate one-tailed statistical significance to compare their scores on a knowledge test on the topic before and after the course, because we know that their knowledge level could have changed in only one direction – up. However, if we wanted to compare two populations of students – one that took one version of the course and another that took a different version, we would then have to calculate the two-tailed statistical significance, because we cannot be sure that the difference between groups is possible in just one direction, even if we had an idea that one version of the course is better. All this said, we should remind ourselves at this point that, in the described example, we would be calculating statistical significance (and using inferential statistics) only if we wanted to know what would the effects of the course be on people in general who are similar to those students in case they took it. If we tested all of the students who took the course and were interested just in their performance, we could compare their results just by looking at the descriptive statistics and comparing them directly without calculating statistical significance or using any inferential statistics at all. As said, before, inferential statistics is used solely when we want to make inferences about a population based on the data from a sample. For drawing conclusions about the sample only, sample data is sufficient!

Since **statistical significance** calculation is based on the sampling distribution, its **values are affected by the same factors that define the width of the confidence intervals – the variability of the data and the sample size** i.e. the number of entities in the sample. The more uniform (i.e. the less diverse) values in the sample are and the larger the sample is, the lower the values of statistical significance i.e. the p values. This means that **the same level of difference from the values proposed by the null**

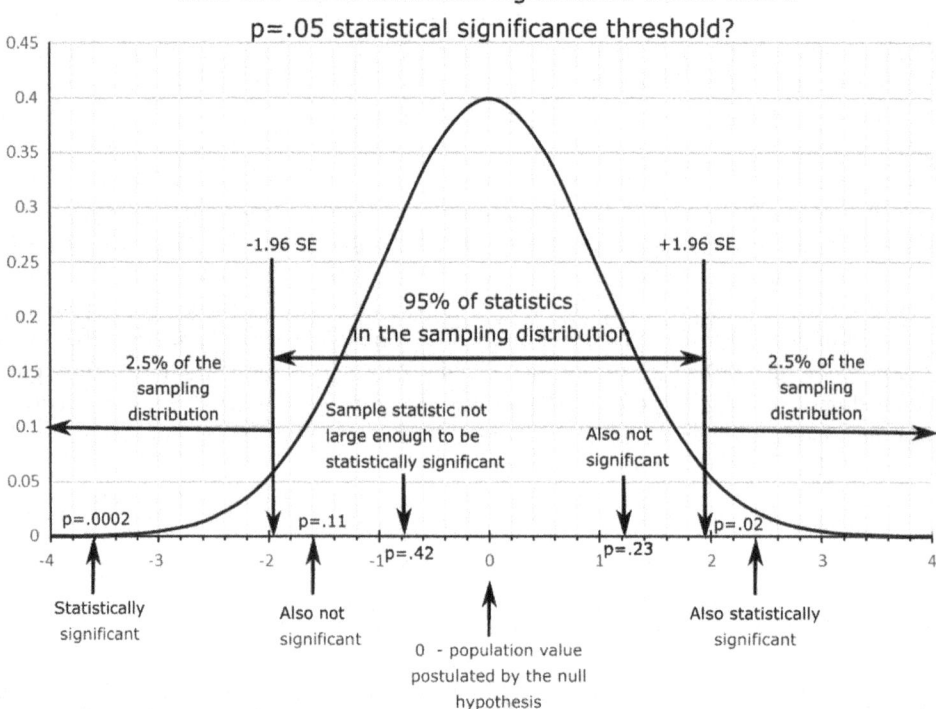

Figure 5.3 A graphical representation of how statistical significance is used to decide whether to accept or reject the null hypothesis in case of using one-tailed significance. This example also uses the .05 statistical significance threshold. Since this is one-tailed statistical significance approach, we have also specified our expectation that differences from the null hypothesis, if they occur, are in the negative direction. That means that all positive differences (contrary to the hypothesized direction) are seen as not statistically significant. Such approach could be validly applied in practice if we knew that positive directions of difference from the null hypothesis in the population would be impossible and must be errors (if differences in both directions are possible, we should use two-tailed significance!!). We can see here again that differences that are not large enough to have p values lower than the accepted threshold are not statistically significant. However, since, this is one-tailed significance, outside of the non-significance area are only 5% of cases on one side of the distribution, while all the cases on the other are included in the non-significance area. This is why the boundary is marked only on one side and it is closer to the mean (1.64 standard errors, than was the case with the two-tailed significance. We can notice also, how sample statistics that have the same distance from the mean as in figure 5.3. have twice lower p values here.

hypothesis (i.e. from 0) will correspond to a smaller p value if the sample size increases and the same will happen if the variability of the variable (from which the data originate) decreases. If these two factors are kept fixed, the main determinant of the value of statistical significance is the difference between the statistics obtained on the sample and the expected values of these statistics according to the null hypothesis. The larger this difference is, the lower the value of statistical significance.

When discussing the values of statistical significance, we should also be aware of a certain contradiction between the way statistical significance is referred to in everyday

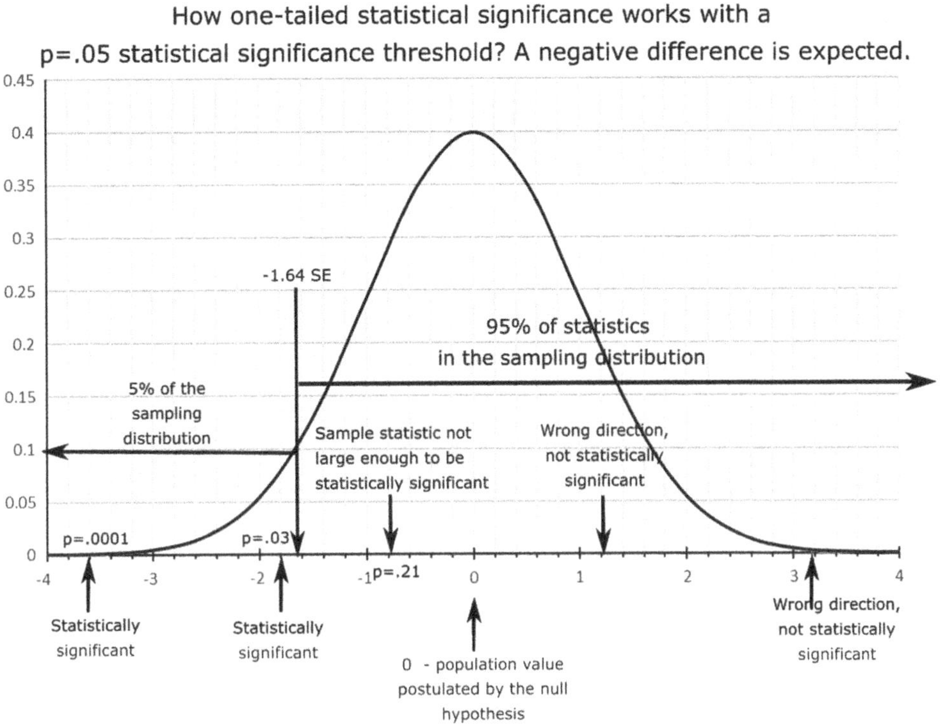

How one-tailed statistical significance works with a
p=.05 statistical significance threshold? A negative difference is expected.

Figure 5.4 A graphical presentation of how statistical significance is used to decide whether to accept or reject the null hypothesis in case of using two-tailed statistical significance. This example uses .05 statistical significance threshold. We can see the 95% confidence interval created around the value postulated by the null hypothesis, which is 0 in this case, and the span of values that are non-significant between –1.96 and 1.96 standard errors. Since this is a sampling distribution, we are considering differences from the null hypothesis in both directions and that means that this threshold leaves outside the non-significance interval 2.5% of statistics on the sampling distribution on one side and the same percentage on the other side of the distribution. We can than see that all the sample statistics (indicated by short arrows and their p values i.e. their statistical significance levels) that are within the confidence interval are not significant, indicating that we should accept the null hypothesis in their cases (all p values above .05), and we also see those outside the confidence intervals (p values lower than .05) that are statistically significant indicating that we should reject the null hypothesis. Please note how the p values decrease with the distance from the center of the sampling distribution created around the value proposed by the null hypothesis in both directions.

scientific statistical jargon and the numerical values of statistical significance. Namely, **in scientific jargon, a "higher level of statistical significance" means lower values of statistical significance**. For example, **scientists will typically say that the statistical significance value of .01 is more statistically significant or that it indicates a higher level of statistical significance than .02**, which again represents a higher level of statistical significance than the value of, for example, .2. This means that, **in the statistical jargon, higher or greater statistical significance will typically mean a lower value of statistical significance**. This peculiarity of the scientific statistical jargon is very important to note, especially for beginners in the area of research statistics, as not understanding it might lead to a lot of confusion when reading scientific publications and following scientific discussions.

Another important thing to be aware of is that by relying on the statistical significance level to decide whether to reject or accept the null hypothesis and selecting a certain critical level as the threshold for accepting/rejecting the null hypothesis, we are also accepting that **our conclusions about whether the null hypothesis holds or not will be wrong in a certain number of cases**. This will happen either because we accepted a wrong null hypothesis or because we rejected a correct null hypothesis. These errors are called statistical errors:

- The error that we make when we decide to **reject the null hypothesis when it is true** is called the **type 1 statistical error**. Type 1 error is also called **the false positive** – declaring that the result (specified by the alternate hypothesis) exists, when it really does not.
- The error that we make when we decide to **accept a null hypothesis that is not true** is called the **type 2 statistical error.** The type 2 statistical error is also called **the false negative** – declaring that the expected effect does not exist, when it really does exist.

The type 1 statistical error happens when we decide, based on statistical significance, that it is not likely that our sample could come from a population in which the null hypothesis is true, because values of the statistics we are considering on the sample are too different from the values specified by the null hypothesis. However, events of low probability sometimes do happen, so if we take 5% or .05 as the threshold of statistical significance, that means that for 5% of samples taken from population (or populations) in which the null hypothesis is true, we will conclude that they do not come from such a population. If we decrease the statistical significance threshold value (i.e. in the statistical jargon – require greater statistical significance for rejecting the null hypothesis) to 1% or .01 than we can expect there to be 1% of situations when we will be rejecting a null hypothesis that is true in the population. So**, increasing the required statistical significance for rejecting the null hypothesis** (i.e. reducing the statistical significance threshold value, requiring the p value to be smaller) **will also reduce the probability of making a type 1 statistical error**. However, at the same time, **the probability of making a type 2 statistical error will be increasing**. This happens because, as said earlier, requiring a lower p value (i.e. greater statistical significance) for rejecting the null hypothesis increases the range of sample values for which we will accept the null hypothesis. This then means that there will be more possible population values different from the null hypothesis that are not sufficiently different from it to have statistics from their samples cross the threshold values often enough (sufficiently often, because even when there are no differences between populations, differences between their samples will sometimes cross the threshold value, remember!). In other words, for researchers to be able to correctly and reliably detect the differences between the real population values and population values expected according to the null hypothesis, the difference between these two needs to be sufficiently large. Small deviations from the null hypothesis are ever harder to detect through the use of statistical significance the smaller they are. Therefore, **making our statistical significance threshold requirements stricter (by requiring a smaller p value for rejecting the null hypothesis) will decrease the likelihood of making a type 1 error, but increase the likelihood of making a type 2 error and vice versa.**

So what do we do about these errors? We should first make it clear that **whatever we do, we can never completely eliminate error, there will always be certain non-zero probability of making an error (i.e. making the wrong conclusion)**. That is why **it is generally not wise to make decisions with important practical consequences just on the basis of results from a single sample**. If the decision we are making based on the results of statistical inferences is important or can have serious consequences, it should be based, whenever possible, on a series of independent studies consisting of a series of independent samples.

The **second thing that we need to consider is what are the costs of type 1 and type 2 errors.** Is it more costly for us to miss an effect that really exists, or to declare that an effect exists when it really does not? These costs depend on the nature of the phenomena we are studying. For example, if we are testing novel ways to treat certain common ailments, ailments for which effective medicines already exist and comparing the effectiveness of these novel procedures to the already existing ones, we would probably like to keep the likelihood of a type 1 error to a minimum. This is because type 1 errors might lead us to a false conclusion that an ineffective procedure is better than the already existing one. We would then invest our money, in significant amounts, in developing this procedure and implementing it in clinical practice. However, it would soon become known that the new procedure is actually not more effective than the existing one (but is, let us say, more expensive). When this becomes known, we would lose clients, the procedure would be abandoned and the investment in it would be forfeit. Therefore, in such situations, it would be much less dangerous to miss a really effective new procedure, thanks to a type 2 error, than to invest into an ineffective procedure thanks to a type 1 error.

On the other hand, if we were an air defense radar operator during a war, and our job was to alert the air defenses when we detect incoming enemy planes or missiles, it would be much more dangerous for us to make a type 2 error i.e. not notice incoming enemy planes, than a type 1 error i.e. raise a false alarm by declaring that there are incoming enemy planes when there are none. Making a type 2 mistake in this situation will likely result in material damage and deaths of our friends, family and colleagues or even our own, while making a type 1 mistake in this situation would likely only result in a few disgruntled and alarmed colleagues.

Therefore, when deciding on the statistical significance threshold, an important consideration to make is which of these types of errors it is more important to avoid. If we decide it is more important to avoid type 1 errors, we should then move the statistical significance threshold towards lower p values. If it is more important to avoid type 2 errors, we should move the threshold towards higher p values.

Table 5.1 Schematical representation of the two types of statistical errors and two types of correct conclusions.

Statistical errors		*Do we think that there is an effect?*	
		Yes	*No*
Is there an effect?	Yes	Great! **Our conclusion is correct!!** We rejected the false null hypothesis!	**Type 2 statistical error.** There is an effect, but we think that there is none. **We accepted the null hypothesis that is not true!**
	No	**Type 1 statistical error.** There is no effect, but we think that there is. **We rejected the null hypothesis that is true!!**	Great! **Our conclusion is correct!!** We accepted the correct null hypothesis

Finally, what we can do **to reduce the likelihood of both types of errors is to narrow the sampling distribution as a whole, to reduce the standard error**. As said before, the relative width of the sampling distribution depends on the variability of the variable we are considering and the sample size. Since variability is what it is i.e. we cannot reduce it without compromising the representativeness of our sample, what we can do is increase the sample size. **Increasing the sample size will reduce the standard error, thereby narrowing the sampling distribution and will thus make smaller differences from the null hypothesis cross the statistical significance threshold, reducing both the occurrence of type 1 and type 2 errors** (although not changing their relative ratios!). So, if you want to reduce the occurrence of statistical errors, you should increase sample size. On the other hand, increasing sample size, also increases the costs of the study. If the increase in sample size was not really needed, then we have wasted money on collecting data from additional entities that we did not really need. However, if it turns out that our sample was too small for the effect that we wanted to study to be reliably detectable (because, remember, the difference of population values from the values proposed by the null hypothesis needs to be of a certain size in order to make it sufficiently likely that sample statistics will cross the statistical significance threshold) and we thus commit a type 2 error (without knowing it, of course), the money and effort invested into the study are again wasted. That is why researchers use a series of procedures called **power analysis** in order **to determine the minimum sample size needed in order to have a certain probability (usually 80% or 90%) of detecting the effect (i.e. difference from the null hypothesis) of a specified expected size while using a certain statistical significance threshold.** This is referred to as the **power of the study and it is typically expressed as a proportion of tests/studies that would yield statistically significant results (i.e. that cross the threshold of statistical significance, meriting the rejection of the null hypothesis) if the expected effect existed in the population.** In this way, a power of .92 would indicate that 92 out of 100 tests would be statistically significant if the expected effect existed in the population. Power analysis procedures typically combine information on the expected effect size, the desired statistical significance threshold and the desired probability that a sample from the population in which the expected effect size exists crosses the statistical significance threshold to calculate the minimum sample size needed. **A typical assumption used in power analysis procedures is that the sample is drawn from the population through random sampling**. It should also be noted here, that **sample sizes typically found in published research in the area of social and behavioral sciences are much, much larger than the minimum sizes that would be indicated by the results of power analysis**. Therefore, power analysis procedures are most useful in situations when it is not possible to collect a large sample, either because entities (or study participants, when the entities are people) are rare or because the measurement procedures are complicated, expensive or extensive. In other words, **power analysis procedures are most useful when we are restricted to using very small samples.** When samples are large, even very small effects easily cross the significance threshold and the problem appears of deciding which of the detected effects (i.e. differences from the null hypothesis) are really worth considering further and which can be neglected because they are too small to be of any practical relevance.

****

Although statistical significance has been the central point of inferential statistics and hypothesis testing in social and behavioral sciences for most of the 20th century and it mostly remained so in the first decades of the 21st century, it has received much criticism which have led to researchers starting to apply alternatives to solely relying on statistical significance to derive conclusions about the results of research studies.

One of the main shortcomings of the concept of statistical significance is the fact that its value depends directly on the sample size. Due to this, when the sample is small, even large deviations from the null hypothesis (i.e. large effects) can remain statistically nonsignificant (not passing the statistical significance threshold) and thereby undiscovered. On the other hand, when the sample size is large, even negligible differences, differences without any practical relevance, become easily statistically significant. The shortcoming is so important that some researchers even pejoratively refer to statistical significance calculation as "sample size testing" or "the null ritual" (Gigerenzer, 2004). This situation then makes a perfect setting for dishonest research approaches – if a dishonest researcher would like to show that a certain effect does not exist, he/she can do a study of that effect on a sample that is too small and then report that the effect was not detected (which would then be interpreted as "does not exist"). On the other hand, if such a researcher wanted to showcase a negligible effect for reasons unrelated to really gaining scientific knowledge, he/she could do a study on a very, very large sample and on such a sample, given multiple comparisons, it is quite likely that something would turn out to cross the statistical significance threshold. Such a researcher could than conveniently "forget" to comment on the size of the difference from the null hypothesis and focus on the fact that a statistically significant effect was obtained.

This last example brings us to the other important shortcoming of the sole reliance on statistical significance to draw conclusions about results of research studies – statistical significance used in the way described above, produces solely a binary decision – to accept or to reject the null hypothesis. However, **rejecting the null hypothesis tells us nothing about the likely size of the difference between the study population and the values proposed by the null hypothesis, but only whether these differences likely exist or not.** And this is sometimes not sufficient, because, as mentioned before, studies on very large samples can make differences that are too small to be of any practical relevance, cross the statistical significance threshold i.e. become statistically significant.

Due to this, it is now **recommended that a measure of effect size** (i.e. the size of the difference from the values proposed by the null hypothesis) **be used alongside statistical significance**. It should be noted that, at the moment this book is written, in 2021, the use of measures of effect size is widely accepted, but there are also calls to abolish the use of statistical significance for drawing conclusions about outcomes of research studies altogether. However, in spite of those calls, the use of statistical significance seems to be still alive and well.

Another line of criticism targets the conceptual basis of null hypothesis testing through statistical significance as described above calling it, somewhat pejoratively, null hypothesis significance testing or NHST (e.g. Perezgonzalez, 2015). They correctly note that the null hypothesis testing procedure most commonly used in social sciences, biology, psychology and other fields i.e. the one described in this book is a combination of two data testing approaches – one proposed by Fischer and the other one proposed by Neyman and Pearson in the first half of the 20th century (Gigerenzer, 2004; Haig, 2017; Lakić, 2019; Perezgonzalez, 2015). They note that, in this approach, some properties of these two theoretical approaches have been combined in a way they consider incongruent and that other things have been greatly simplified starting with the null hypothesis, which has, more or less, been reduced to a hypothesis stating that parameter value is zero, that alternative hypotheses are often not formulated, that researchers using this approach often regard it as a ritual without really understanding the concepts the procedures are based on, that the most commonly used critical levels of statistical significance (mostly .05) are, in practice and in teaching students and beginner researchers, treated completely inflexibly as rigid thresholds, instead of more or less arbitrary cut points and that many things researchers do are plainly wrong (such as, for example, interpreting nonsignificant results as "a tendency" towards the state of affairs that would give merit to the rejection of the null hypothesis). A rather comprehensive overview of the similarities and differences between the original Fischer's and Neyman-Pearson approaches on the one hand and the "null hypothesis significance testing" approach found in research practice can be found in Perezgonzalez (2015), who presents a detailed overview of all three approaches and a table detailing the similarities and differences between them in key points. While many points of this criticism are valid on the conceptual level, we can ask ourselves whether they really disqualify the null hypothesis significance testing as an approach that is "good enough" for practical use in research. The answer to that question, in our opinion, is a resounding no. While, indeed, the described procedures have many shortcomings, most important of which have already been mentioned in this book, they are also responsible for introducing and spreading the use of statistical and other mathematical techniques to areas of science where they were previously much less present, thus changing the whole methodological concept of research in these areas. Its comparative simplicity allowed for its feasible integration into study programs in scientific areas that previously had practically no mathematics and allowed scientists in areas traditionally not very close to mathematics the adoption and use of relatively sophisticated statistical techniques in their research. It would not be an exaggeration to say that introduction of statistical procedures starting with the ones described here, revolutionized the way research was done in social sciences, psychology, but also biology and some other scientific areas. This can best be seen when comparing the scientific quality and value of research done by researchers in these scientific areas coming from parts of the world where this integration of statistics into the research methodology happened decades or over a century ago with research done by researchers coming from environments where this did not happen at all or the start of this integration is of a relatively recent date. While there are voices that are of the opinion that events such as the replicability crisis in psychology (the finding of a recent series of studies replicating classical studies in psychology that did not manage to replicate much of the expected results and found many of those that were replicated to produce lower effects than originally reported) (e.g. Świątkowski & Dompnier, 2017) is partly due to the shortcomings of the application of inferential statistics in psychological research and of the understanding of statistical concepts by psychologists (Lakić, 2019), we firmly believe that without the simplifications introduced in the practical applications of statistics done by researchers, there would be no widespread introduction of statistics into research in these areas and without it there would be no replicability. To have a crisis of replicability, there first need to be research results that are replicable and these would not have existed had the researchers not adopted statistics. Also, to be able to replicate a study and obtain effects of lower intensity than those reported in original studies, there first have to be original studies reporting effects, which again would not be there without statistics. So, in spite of the existing and potential shortcomings of the statistical procedures currently in widespread practical use, the fact that their introduction has revolutionized and greatly advanced the quality of research in many scientific areas is undeniable. That said, statistical methods are developing and room for improvement definitely exists. Some of the possible improvements will be presented here. Also, some of the criticisms directed at these statistical procedures are plainly unjustified. For example, the criticism about the rigidness of the procedures applied in scientific research, the claims that scientists are required to all apply the same procedures and decision-making rules including the same threshold statistical significance level, pejoratively referred to by some critics as "the null ritual" or the mindless use of statistics (Gigerenzer, 2004), actually represent the desirable traits of standardization of scientific procedures and promotion of objectivity. Standardized scientific procedures and objectivity are traits of particular importance in science in general, but especially in areas such as social sciences, areas with a long history of being heavily influenced or even completely hijacked by political ideologies and agendas, undisclosed individual beliefs and general subjectivism to such an extent that has often derailed scientific progress and nullified the value of such research (e.g. Hedrih, 2020; Reyna, 2017). In such an environment, standardization of research procedures in a way that makes sure that all researchers will draw the same conclusions from the same set of data and the same set of statistical procedures is an important advantage, not a shortcoming, even if it comes at the price of somewhat increasing the theoretical inconsistencies above the preexisting level. This should of course not be taken to mean that there is no room for improving the null hypothesis testing procedures in common use. On the contrary, and many such improvements have already become part of mainstream scientific practice.

Testing the null hypothesis using resampling. Procedures for calculating statistical significance have traditionally relied on the central limit theorem, meaning that the sampling distribution properties were estimated using formulas, as described in the previous chapter. An alternative to this, coming at the moment this book is written into common use in various statistical software packages, is the testing of null hypotheses through the use of bootstrapping. The null hypothesis testing through bootstrapping works in a way similar to how confidence intervals for estimating parameter values are created:

- In the first step, the **desired number of samples are drawn with replacement from the original study sample** and, on each of them, **calculations of statistics for which the null hypothesis is tested are performed.**
- A **sampling distribution is created** from these statistics.
- **Boundary points for the confidence interval on the sampling distribution are defined based on the desired threshold** i.e. based on how large percentage of the distribution we wish to encompass by the confidence interval. **The most typically used confidence intervals in literature are 95% and 99%.** These can be taken to correspond with statistical significance levels of 5% and 1% i.e. .05 and .01, because when 95% of the sampling distribution is included in the confidence interval, 5% of the distribution are left out of the interval and this proportion of entities from the sampling distribution outside the confidence interval is essentially what statistical significance represents. It should be noted that this confidence interval is formed in such a way that its upper and lower boundary are equally distant

from the sampling distribution mean. This confidence interval is usually referred to as the **bootstrap confidence interval** with **the percentage of the sampling distribution included in it also used in the name**. For example, if a bootstrap confidence interval is created so that it includes 95% of entities from the sampling distribution it is called a 95% bootstrap confidence interval, if it is created to include 99% it is then called the 99% bootstrap confidence interval and so on. The procedure typically also calculates the difference between the mean of the sampling distribution and the mean of the study sample (the actual sample collected empirically from which the bootstrap samples are drawn) and this value is called **bias.**

- The **null hypothesis is than evaluated by noting whether the value specified by the null hypothesis (that value is most typically 0) lies within the confidence interval created in the previous step or outside of it.** If the **value specified by the null hypothesis lies within the confidence interval then the null hypothesis is accepted.** If the **value specified by the null hypothesis lies outside the confidence interval, then the null hypothesis is rejected**.

For example, if we wanted to test the null hypothesis that two samples come from populations that have the same mean (on some variable that was assessed), we would create the null hypothesis that the difference between means of the two populations on this variable is 0. We could than perform bootstrapping and create, for example, 10 000 pairs of samples from this original pair of samples and then calculate the difference between means in each of the 10 000 pairs of samples. We would obtain 10 000 differences between means in this way and these would constitute the sampling distribution of the difference between means. We would than define, let us say, a 95% bootstrap confidence interval, such that the lower and upper boundary of the interval equally differ from the mean of the sampling distribution. Let us suppose, that we created this interval and that it ranged, for example, from −1 to 3, with the mean of the sampling distribution being 1. Since our null hypothesis specifies that the difference between population means is 0, we would than inspect the confidence interval we obtained and conclude that 0 lies within the confidence interval. Because of this, we would conclude that the null hypothesis holds and we would accept it. If, however, the confidence interval turned out to be between, for example −3 and −1 or between 2 and 5, we could than conclude that 0 (the value specified by the null hypothesis) lies outside the bootstrap confidence interval and thus reject the null hypothesis. It should be noted here that, although null hypotheses specifying the value of zero (i.e. nil hypotheses) are practically exclusively found both in literature and statistical software, any specific statistical hypothesis (i.e. specifying a certain precise value of parameters) can equally well serve as a null hypothesis.

As with using bootstrapping procedures to estimate values of parameters, **the advantage of bootstrapping in comparison to calculating statistical significance based on the central limit theorem is that it does not rely on assumptions about the properties of the sampling distribution,** but directly calculates its properties from a sampling distribution created by drawing samples out of the original study sample.

5.5 Bayes' factor and testing statistical hypotheses

One important property of testing null hypotheses through calculating statistical significance and also one of the main points of criticism is the fact that, in the way it is typically used, statistical significance testing makes no use of the existing theoretical knowledge about the matter being investigated (e.g. Dienes, 2014; Dienes & Mclatchie, 2018; Gigerenzer, 2004; Lakić, 2019). The researcher specifies the null hypothesis, the one that is expected to be false, but there is no specification (in the statistical procedure) of what the researcher actually expects. While this approach is certainly valid when there is no previous knowledge of the phenomenon studied and thus no basis for having expectations, with phenomena that are well studied researchers often have basis to formulate and test specific hypothesis, yet this is rarely done. Also, when the statistical significance level indicates that we should accept the null hypothesis, this does not really mean that the hypothesis is likely to be correct, rather only that it could not be rejected based on the procedure applied. To overcome this problem, some researchers propose that instead of calculating the statistical significance of the null hypothesis, **we should be comparing competing hypotheses and one way to do that is by calculating the Bayes factor**. The procedure for calculating the Bayes factor is a part of an approach in statistics called Bayesian statistics, which is a theory based on **the Bayes interpretation of probability, where probability is seen as an expression of the degree of belief in an event**. Bayesian statistics is named after the 18th century English statistician Thomas Bayes, who authored what is now called the Bayes' theorem. The field was later developed further through works of the French mathematician Pierre-Simon Laplace.

This particular approach to testing statistical hypotheses starts by formulating the competing hypotheses, one of which can be considered the null hypothesis and the other one the alternative hypothesis, but these can also be two competing hypotheses based on differing results of previous research. These hypotheses propose that the parameter has a certain value or that it can have a certain range of values. Based on that, a distribution of probabilities of obtaining different values of statistics on samples for each hypothesis is created by researchers or through certain statistical procedures. This is done before the data is collected. These probabilities are called **a priori probabilities**. After the data collection, **the researcher calculates the exact probability of obtaining the result that was obtained on the study sample in case the first hypothesis is true and then in the case the second hypothesis is true (in the population). Dividing the first probability** (the probability of obtaining the result on the sample as was obtained when the first hypothesis is true) **with the second probability** (the probability of obtaining the result on the sample as was obtained when the second hypothesis is true) **results in what is called the Bayes factor, typically denoted as BF.** The value of the Bayes factor of 1 denotes that both hypotheses are equally likely given the data at hand. Values of the Bayes factor above 1 indicate that the first hypothesis (the one that constitutes the dividend in the ratio) is more likely, while values below 1 indicate that the second hypothesis is more likely (the one that constitutes the divisor in the ratio). Considering the interpretation threshold, a commonly found recommendation is that the **BF values above 1 and up to 3 be interpreted as inconclusive** (about which of the two hypotheses is more likely), **values 3–10 can be seen as moderately strong evidence in favor of the first hypothesis**, such as would be expected in early phases of research, **10–30 are seen as strong evidence in favor of the first hypothesis, above 30 and up to 100 as very strong and BF values over**

100 as extremely strong, crucial evidence in favor of hypothesis 1. In a similar fashion, **values below 1 and down to 1/3 (or .33) are seen as inconclusive, between .33 and .1 as moderate evidence in favor of hypothesis 2** (the one constituting the divisor in the Bayes factor ratio), **values between .1 and 0.033 as strong evidence, between .033 and .01 as very strong evidence and below .01 as extremely strong or crucial evidence in favor of hypothesis 2** (Lakić, 2019; Nicenboim et al., 2021).

A key property of this approach is the formulation of the hypotheses i.e. the definition of a priori probabilities. This can be done in various ways, from declaring that all possible values of the variable have equal probability (thus creating a uniform a priori distribution of probabilities), to very precise probability distributions based on extensive previous knowledge. This, by its very nature, requires the reliance on the judgement of the researchers to formulate hypotheses and therefore can possibly create situations where different researchers would formulate them in different ways given the same data available prior to research, meaning that it is subjective. As Lakić notes in his paper on the Bayes factor (Lakić, 2019), calculation of the Bayes factor can also be done in an objective way, by, for example, assigning equal probabilities to all possible values of the considered statistic. However it can be said that doing so negates the primary idea behind this approach and that is the inclusion of previous knowledge on the topic under study in the calculations. This necessary subjectivity in the approach could also, potentially, be abused by researchers keen on demonstrating that their hypothesis was correct, by formulating their hypotheses after the data has already been collected (and they had a look at it) and not before that, as is required. However, the approach using the Bayes' factor sees research as a dynamic process not as a one-off event. This means that even if a researcher did this, future researchers and studies, considering the same collections of previous data would easily bring the validity of these findings into question. One important thing to note is that the Bayes factor alone cannot tell us which hypothesis is the most probable, but rather tells us given the study data and the prior data available, by how much we need to update our relative belief between the two competing hypotheses (Nicenboim et al., 2021).

Having all this in mind, **the main property of the Bayesian approach – its reliance on the judgement of researchers estimating a priori probabilities seems to be its main shortcoming,** as this also makes this approach much more complex and difficult for researchers not too familiar with statistics. This is likely one of the reasons for its still relatively modest level of use in the current published research in social and behavioral sciences.

5.6 Parametric and nonparametric statistics

At this point, it is useful to draw a distinction between two types of statistical procedures that differ in whether they make the assumption about the distribution of data or not. **The statistical procedures that make assumptions about the properties of the distribution of data and also possibly other assumptions are called parametric procedures. Those that do not are called nonparametric procedures.**

Parametric procedures most commonly found in scientific literature in the area of social and behavioral sciences rely on the assumption that something is normally distributed. That might be data or the sampling distribution, but, in more advanced statistical procedures, this assumption may refer to distributions of other things. Aside from this, these procedures usually also rely on the following assumptions:

- **that the data are at least on the interval level of measurement** – this goes together with the expectation of a normal distribution. The normal distribution is a distribution of data that is at least interval, it relies on the existence of the fixed unit of measurement. Without it, there would be no way to establish whether the shape of the distribution is normal or not i.e. whether the values are grouped as expected according to the normal distribution. It would also be impossible to calculate the descriptive statistics for normally distributed data – the mean and the standard deviation as both of these require at least interval data. So, if the data are ordinal or nominal, we cannot have a normal distribution.

- **that the sample data come from independent observations/measurements**. This basically means that values of any one entity did not influence the values of any other entity and that different pieces of data come from different measurements and there are no multiple copies of the same measurement presented as different measurements. In the case of social and behavioral sciences, with people as entities from which data were collected and data being their answers to various questions or test stimuli, this most commonly means that study participants answered individually and independently without conferring with one another about what to answer and also without being instructed how to answer.

- **That variances are homogenous across the considered groups** – this assumption essentially means that differences between groups, if they exist, are in the mean expression levels of the measured variables and not in the variability of these variables. This also implies that when we are testing for the effects of various factors, we assume that these factors affect all entities in a roughly equal way, thereby creating differences between means of groups that were exposed and those that were not exposed to those factors and not in their variances. If we are considering multiple groups, this assumption means that all groups have the same variance. If we are considering relationships between variables, this assumption means that variance of one variable is the same on all levels of the other variable.

It should be noted that **some parametric tests also have variants that do not assume equal variances across data** and when these variants are calculated it is customary to note that in the presentation of results.

When discussing statistical procedures in the later part of this book, it will be noted whether the procedure is parametric or nonparametric. For most data analysis needs, there exist both parametric and nonparametric procedures. For example, the procedures for estimating population parameters from sample data based on the central limit theorem, rely on equations for estimating standard errors that in turn rely on the assumption that the sampling distribution is normal and are thus parametric procedures. On the other hand, the bootstrap procedure for estimating parameters does not really on such assumptions and is therefore nonparametric.

5.7 Let us apply what we learned so far!

Let us now try to apply what we have covered in this chapter through a couple of exercises. Please refer to the start of the book for the general instruction for completing the exercises. Our suggestion is that you first read the excerpt and the statements and provide your own answer. You may write it in the Answer column, and after that look up the answers and compare your own answers with them.

Exercise I. Parameter estimation, testing null hypotheses. (Hedrih et al., 2017)

Type of vocational interest		Mean	SD	t statistic	Statistical significance
Real type (R)	**Athletes**	2.83	1.19	1.41	0.16
	Non-athletes	2.71	1.26		
Investigation type (I)	**Athletes**	3.49.	1.17.	−0.68.	0.50
	Non-athletes	3.55.	1.23.		
Artistic type (A)	**Athletes**	3.35.	1.52.	−2.71.	0.01
	Non-athletes	3.63.	1.57.		
Social type (S)	**Athletes**	4.04.	1.12.	−3.44.	0.00
	Non-athletes	4.29.	1.08.		
Enterprising type (E)	**Athletes**	3.85.	1.08.	−0.15.	0.88
	Non-athletes	3.86.	1.03.		
Conventional type (C)	**Athletes**	3.20.	1.10.	1.58.	0.11
	Non-athletes	3.08.	1.14.		

The null hypothesis tested in these procedures is that the two samples come from populations with equal means!! The procedure used is parametric.

Please use the .05 statistical significance threshold when answering statements!

The table presents the means and standard deviations of two groups of people (athletes and non-athletes) on a series of vocational interest variables. A statistical test was conducted to test the null hypothesis that means of the populations the two groups come from are equal and the statistical significance is reported in the final column.

Table is based on data from the study of Hedrih et al. (2017).

I	*Statement:*	*Answer*
I1.	We should reject the null hypothesis of this procedure for variable R.	
I2.	We should reject the null hypothesis of this procedure for variable I.	
I3.	We should reject the null hypothesis of this procedure for variable A.	
I4.	We should reject the null hypothesis of this procedure for variable S.	
I5.	We should reject the null hypothesis of this procedure for variable E.	
I6.	We should reject the null hypothesis of this procedure for variable C.	
I7.	The upper boundary of the 95% confidence interval of the mean of variable R on the group of athletes would be higher than 6.	
I8.	The lower boundary of the 95% confidence interval of the mean of variable I on the group of athletes would be lower than 8.	
I9.	Skewness of distributions of both groups on variable R is lower than 0.5	
I10.	Participating in sports causes the reduction of both artistic and social types of vocational interests and that is the reason why there are statistically significant differences between the means of the two groups on these variables.	

Exercise J. Parameter estimation, testing null hypotheses.

Descriptives

			Statistic	Std. Error
p39	Mean		3.93	.084
	95% Confidence Interval for Mean	Lower Bound	3.76	
		Upper Bound	4.09	
	5% Trimmed Mean		4.03	
	Median		4.00	
	Variance		1.691	
	Std. Deviation		1.300	
	Minimum		1	
	Maximum		5	
	Range		4	
	Interquartile Range		2	
	Skewness		-1.198	.157
	Kurtosis		.336	.312

The table presents the descriptive statistics and standard errors of an answer to a questionnaire item from authors' own data.

J	Statement	Answer
J1.	There is a 95% chance that the value of the population mean on this sample is in the range between 3.76 and 4.09.	
J2.	The standard error of the mean of variable p39 is lower than .1.	
J3.	The standard error of standard deviation of the variable p39 is lower than 1.	
J4.	The variance of the variable p39 is higher than 2.	
J5.	The mean of females on the variable p39 is higher than the mean of males.	
J6.	The percentile deviation of the variable p39 is 4.	
J7.	The quartile middle range of the variable p39 is 2.	
J8.	The upper boundary of the 99% confidence interval of the mean of the variable p39 is lower than 4.	
J9.	P39 has a negatively asymmetrical distribution.	
J10.	The 50th percentile of the variable p39 is 4.	

Exercise K. Parameter estimation, testing null hypotheses. (Hedrih, 2017)

Statistic		Sample statistic value	Bias	Standard error	95% confidence interval lower boundary	95% confidence interval upper boundary
				Bootstrap results, 1000 samples		
WFC	Mean	2.62	.00	.04	2.54	2.69
	Median	2.4	.06	.09	2.4	2.6
	Standard deviation	1.19	.00	.02	1.15	1.23
	Skewness	.30	.00	.05	.21	.40
	Kurtosis	−.97	.00	.06	−1.08	−.85
FWC	Mean	1.72	.00	.03	1.67	1.77
	Median	1.4	.02	.06	1.4	1.6
	Standard deviation	.83	.00	.02	.78	.87
	Skewness	1.27	.00	.08	1.12	1.41
	Kurtosis	1.16	.00	.03	.59	1.77

The table presents descriptive statistics and confidence intervals of work–family conflict (WFC) and family–work conflict (FWC) based on a part of the data from Hedrih (2017).

K	Statement	Answer
K1.	The standard deviation of the sampling distribution of the median of WFC is .06.	
K2.	There is no difference between the mean of the sample on the variable WFC and the mean of the sampling distribution of the mean of this variable obtained through the bootstrapping procedure.	
K3.	It is highly possible that kurtosis value of the population from which the sample was taken on WFC is 0.	
K4.	The grand mean quartile is larger for WFC than for FWC.	
K5.	There are more than 500 entities in the sample.	
K6	It is highly probable that the distribution of FWC in the population the sample is from is positively skewed.	
K7.	We can expect that the mean of WFC will be between 2.54 and 2.69 in 95% of repeated studies on samples from the population this sample is from.	
K8.	There is no chance that in any of the future studies on samples from this population the mean of FWC be lower than 1.7.	
K9.	The distribution of the population this sample is from on WFC is likely platykurtic.	
K10.	The procedure presented here for making inferences about parameter values is based on the central limit theorem.	

Let us now consider the answers:

I1 – false. We can see that the statistical significance level is .16. This is higher than our statistical significance threshold of .05 indicating that we should not reject the null hypothesis.

I2 – false. The same reasoning as with I1. The statistical significance level here is .5, which is higher than .05.

I3 – true. The statistical significance level of .01 is lower than the .05, meaning that we should reject the null hypothesis. In the common scientific jargon, we would say that the result is "more statistically significant" than the accepted threshold thus meriting the rejection of the null hypothesis.

I4 – true. The same reasoning as with I4 and the results of this procedure are even "more statistically significant" than was the case with I3.

I5 – false. Not statistically significant. Statistical significance of .88 is far above the .05 threshold.

I6 – false. Statistical significance of .11 is still not enough to reject the null hypothesis given the .05 threshold.

I7 – false. While we do not have confidence intervals of the mean presented here, we know that the upper boundary of the confidence interval is established by adding roughly 2 standard errors (1.96 to be exact!) to the value of the mean. We also know that the standard error of the mean is obtained by dividing the standard deviation with the square root of the number of entities in the sample. This means that the standard error of the mean must be lower than the standard deviation. We can also see that the mean of the group of Athletes on this variable is 2.83, while the standard deviation is roughly 1.2 (if we want to be exact, it is 1.19). Two standard deviations here are roughly 2.4. If we add 2.4 to 2.83 the result is definitely lower than 6, meaning that it would not be possible for the upper boundary of the 95% confidence interval of the mean to be higher than 6.

I8 – true. We can see that the mean of the group of athletes on the variable I is 3.49. The lower boundary must be lower than the sample mean and since the sample mean is lower than 8, so must the lower boundary of the confidence interval of the mean also be lower than 8.

I9 – unknown. There is no data in the table that would allow us to make inferences about the skewness level. Yes, the procedure used (will be covered in the later part of the book) is a parametric one, requiring a normal distribution, but the size of the proposed skewness statistic is still within the range in which the use of parametric procedures is considered acceptable (does not indicate a deviation from the shape of the normal distribution that is sizeable enough to prevent the use of parametric procedures).

I10 – unknown. The data in the table just shows that there are differences between means of the two groups on these two variables and that we would be justified in concluding that the difference in means also exists between the means of the population these two groups were sampled from. However, there is no data on the causes of these differences. It might indeed be that doing sports changes vocational interests, but it might also be that people with different vocational interests differ in their preference for participating is sports. It is also possible that there are other factors that cause differences between people both in vocational interests and in doing sports. Data presented in the table do not allow us to tell which of these options is true, if any. Therefore, we do not know whether the statement is true and we cannot make any judgement about its veracity based on the data presented in the table.

J1 – true. The statement describes what is the 95% confidence interval of the mean. This interval is presented in the table and the range is correct, as can be seen from the table.

J2 – true. We can see that the standard error of the mean is .084, which is lower than .1.

J3 – true. We do not have the standard error of the standard deviation here, but have the standard deviation of the mean. And we know from this chapter that the standard error of the standard deviation is always lower than the standard deviation of the mean. Given this, if the standard error of the mean is .084, which is much lower than 1, so must also be the case with the standard error of the standard deviation.

J4 – false. We can read from the table that variance is exactly 1.691, which is not higher than 2.

J5 – unknown. We have no data in the table to make any conclusions about statistics of males and females.

J6 – meaningless. What exactly might be the "percentile deviation"? There is no such thing to our knowledge.

J7 – meaningless. What exactly might be the "quartile middle range"? There is no such thing to the best of our knowledge.

J8 – false. We can see that the upper boundary of the 95% confidence interval is higher than 4. The upper boundary of the 99% confidence interval must be even higher (look at the formula for calculating these confidence intervals) and hence cannot be lower than 4.

J9 – true. We can see that the skewness value is negative and quite substantial, indicating a negatively asymmetrical distribution.

J10 – true. 50th percentile is the median and we can read from the table that the median indeed is 4.

K1 – false. The standard deviation of the sampling distribution is called standard error, and we can read that it is .09 for the median of WFC. The value of .06 is bias, which is something else.

K2 – true. The difference referred to by this statement is called bias and we can indeed see that it is 0.

K3 – false. We can see that the kurtosis value of the sample −.97 and that the confidence interval of kurtosis presented in the table also does not include the value of 0, making it not likely that that is the population value. Actually, we can see that the value of 0 is over 15 standard errors away from the sample value. This makes the value of 0 in the population very, very, very, very unlikely. This is, of course, on the premise that we are working with a random sample from the population we are making inferences about or a sufficiently good approximation of one, which is a general premise of all inferential procedures presented in this book.

K4 – meaningless. There is no standard statistic called "the grand mean quartile".

K5 – unknown. There is no data on the population size in the table. We could possibly make some inferences based on the relationship between the standard error and the standard deviation if we assumed that the bootstrap standard error value would be similar to the standard error value based on the central limit theorem, but this is in no way certain and such a calculation is too complex to expect from a general reader of a scientific text. It is much easier to ask the authors or to look it up in some other part of the paper (when we are reading full text scientific papers of course, not just a single table for a statistical exercise!)

K6 – true. We can see that the skewness is positive and that also the entire range of the confidence interval is in the area of positive skewness values and this is an interval for which there is a 95% chance that it encompasses the parameter. Also, we can see that the lower boundary of the interval is very, very far from 0. A positive skewness means that the distribution is positively skewed.

K7 – true. Yes, the statement refers to another way to interpret the confidence interval of the mean and the values are correct.

K8 – false. The sample mean is 1.72 and we can see that the value of 1.7 is inside the confidence interval. That means that we can fully expect a certain number of future studies to have the mean of 1.7 or lower. But even if this was not the case, the statement would still be false, as any results, whatever they are, could make it only unlikely to obtain a mean of that value, but not impossible, given the fact that the normal distribution asymptotically approaches 0, never reaching it.

K9 – true. We can see that the kurtosis value is highly negative, indicating a platykurtic distribution.

K10 – false. No, this is a bootstrapping procedure for making inferences about population values, not the procedure based on the central limit theorem.

References

Dienes, Z. (2014). Using Bayes to Get the Most Out of Non-significant Results. *Frontiers in Psychology*, *0*, 781. 10.3389/FPSYG.2014.00781

Dienes, Z., & Mclatchie, N. (2018). Four Reasons To Prefer Bayesian Analyses Over Significance Testing. *Psychonomic Bulletin & Review*, *25*(1), 207–218. 10.3758/S13423-017-1266-Z

Efron, B. (1981). Nonparametric Estimates Of Standard Error: The Jackknife, The Bootstrap And Other Methods. *Biometrika*, *68*(3), 589–599.

Gigerenzer, G. (2004). Mindless statistics. *The Journal of Socio-Economics*, *33*, 587–606. 10.1016/j.socec.2004.09.033

Haig, B. D. (2017). Tests of Statistical Significance Made Sound. *Educational and Psychological Measurement*, *77*(3), 489–506. 10.1177/0013164416667981

Harding, B., Tremblay, C., & Cousineau, D. (2014). Standard Errors: A Review And Evaluation Of Standard Error Estimators Using Monte Carlo Simulations. *The Quantitative Methods for Psychology*, *10*(2), 107–123. 10.20982/tqmp.10.2.p107

Hedrih, V. (Ed.). (2017). *Work and Family Relations at the Beginning of the 21st Century*. Filozofski fakultet, Niš.

Hedrih, V. (2020). *Adapting Psychological Tests and Measurement Instruments for Cross-Cultural Research: An Introduction (1st Edition)*. Routledge, Taylor&Francis Group.

Hedrih, V., Ristic, M., & Randjelovic, K. (2017). Vocational Interests of Recreational Athletes. *Facta Universitatis: Series Physical Education and Sports*, *15*(1), 37–48. 10.22190/FUPES1701037H

Lakić, S. (2019). Bayesov faktor: Opis i razlozi za upotrebu u psihološkim istraživanjima[Bayes Factor: What It Is and Why it Should it be Used in Psychological Research]. *Godišnjak Za Psihologiju*, *16*, 39–58.

Nicenboim, B., Schad, D., & Vasishth, S. (2021). *An Introduction to Bayesian Data Analysis for Cognitive Science*. Bookdown. https://vasishth.github.io/bayescogsci/book/

Perezgonzalez, J. D. (2015). Fisher, Neyman-Pearson or NHST? A tutorial for teaching data testing. *Front Psychol*. 10.3389/fpsyg.2015.00223

Reyna, C. (2017). Scale Creation, Use, and Misuse: How Politics Undermines Measurement. In *The Politics of Social Psychology* (pp. 79–98). Psychology Press. 10.4324/9781315112619-6

Świątkowski, W., & Dompnier, B. (2017). Replicability Crisis in Social Psychology: Looking at the Past to Find New Pathways for the Future. *International Review of Social Psychology*, *30*(1), 111–124. 10.5334/irsp.66

6 Correlations

6.1 Associations between variables

One of the first inquiries to make when starting research in a new area or on a new topic is to identify the key variables and then conduct observations to establish hypotheses about their relationships. An important part of this is identifying which variables tend to change together and which do not i.e. which variables are associated with one another.

In statistics, **a statistic indicating the strength of association between variables is called a correlation**. To declare that two variables are associated, we need to be able to observe that changes in values of one variable are followed, with more or less precision, with changes in values of the other variable. This joint change of values is a matter of a degree – while there are examples when changes in one variable are perfectly matched by corresponding changes in the other variable, most of the times in practice, these joint changes will be less precise, up to a point where there is only a very loose association between the two variables.

This said, we should be aware that **correlation i.e. association between variables does not imply a causal relationship! If two variables are correlated that does not mean that one causes the other** (i.e. that changes in one variable cause the changes in the other variable). **If one variable causes the other i.e. if there is a causal relationship between two variables, than they will likely be correlated, but variables can be correlated/associated for a whole number of different reasons and causation is only one of them. It is also possible that both variables are influenced by a third variable causing them and that this is the reason for the observed correlation.**

The **nature of the relationship between two variables can also be much more complex**, with both variables being a part of the same causal chain or network, where multiple other variables influence both of them leading to the correlation. **In situations** like this, **where variables are correlated, but not causally related (one does not cause the other), deliberately changing the values of one variable will not lead to the changes in the other variable and these induced changes will actually lower the level of association between the two variables or completely nullify it. Changing the values of one variable will also not affect the other variable in the situation when the variables are causally related, but the one we are changing is the consequence of the other. The only situation when changing the values of one variable will lead to changes in the other variable is when the variable we are changing is the cause and the other variable is the consequence** (or somewhere down the line of consequences of the first variable) and that is only a part of all correlations between variables found in scientific research. **It is very important for readers to always remember that correlation is not causation,** although some correlation relationships

DOI: 10.4324/9781003107712-6

may be due to a causal relationship between the variables. This emphasis is important, because the history of science and particularly of the application of science is littered with situations where simple correlation relationships were misinterpreted as causal relationships and when, due to this misinterpretation, people wrongly believed that changing one variable from a known correlation relationship will cause changes in the other. Some of the historical examples for this were mentioned in the chapter about self-fulfilling prophecies and the corruption of statistical indicators and another very common mistake worth mentioning is the historical tendency to describe whatever foods or behavioral habits are favored by the wealthier parts of the society at the time as healthy or useful in improving one's health, based on observed correlations between the consumption of these foods or practice of such habits and various health indicators alone. While there likely are behavioral habits and food types that can help improve one's health, oftentimes these associations were simply due to the fact that wealthier parts of the population had better access both to healthcare (leading to better health!) and those exotic foods/behavioral practices due to their better financial standing, thus creating a correlation. However, people who did not have access to equal financial resources and thus the quality of healthcare, but just adopted the proposed habits of the wealthy, soon found that the expected health effects do not follow. There is also the very simple example found in many introductory textbooks in statistics – if we calculated the correlation between the number of teachers and the number of thieves across a number of cities, we would see that there is a very strong association between the number of thieves in a city and the number of teachers. Based on this, a person not differentiating between correlation and a causal relationship might conclude that teachers somehow attract or produce thieves (or vice versa?). However, the reason for this correlation is simply that there is a third variable responsible for the correlation – city size. Larger cities will have more people and that both means more teachers and more thieves and actually more of people of any other profession.

Another property of the correlation relationship is that, **without additional theoretical knowledge, the existence of a correlation between variables does not imply that such a correlation will exist in the future.** As was discussed in the first chapter, in the part about statistical explanations and also about the corruption of statistical indicators, **to know whether an association between variables will continue into the future, it is necessary to know why such an association exists in the first place** i.e. we must have a valid and comprehensive scientific explanation of the association between them (ideally, a causal one, although often other kinds of explanations might do). Without such an explanation, a past correlation might simply cease to exist anytime in the future or a correlation might appear in the future where there was none in the past. And this newly found correlation might disappear again later or might change in intensity. This is a phenomenon particularly well known to professionals working in predicting prices of goods and particularly to those involved in predicting the prices of financial instruments on organized financial markets such as stock exchanges. As stock market traders are accustomed to saying – "past gains are poor indicators of future gains" and the same goes for correlations. This is especially the case when elements of human behavior represent the variables in question.

6.2 Types of associations between variables

What can associations between variables look like? While we often think of associations between variables as simple – when one increases, the other one increases also, in practice associations can take many different forms, from very simple relationships to very complex ones. For example, if we observe two dancers performing a complex choreography together, we could conclude that their movements (during the dance) are related even when they are completely different. If we knew how the choreography looked like and knew which movements is one dancer performing at the moment, we could very precisely predict the movements of the other dancer at the same moment, even if we were not able to see what he/she is doing. On the other side of the association complexity come relatively simple relationships like for example the relationship between the level of pressure on the gas pedal of a car and speed of the car (on a level surface, in the same gear...) or between various similar personality traits such as the association between Openness to experience and artistic vocational interests (e.g. Hedrih, 2009). In between these are a bit more complex relationships such as, for example, the relationship between a person's estimate of own competence in an area and actual competence as described by the famous Dunning-Kruger effect (in this study, people whose scores were in the bottom quartile on various tests the researchers administered, overestimated their real scores much more than those who were in the higher quartiles, with the top performers underestimating their performance) (Kruger & Dunning, 1999) or the known logarithmic relationship between the physical intensity of visual or sound stimuli and their perceived intensity, also known as the Weber-Fechner law (e.g. Portugal & Svaiter, 2011).

When considering the association between variables an important concept to understand is the concept of a direction of association between variables. **A direction of association between variables refers to the direction in which values of one variable change when values of the other variable change**. When **increase in the values of one variable is accompanied by an increase in values of the other variable this is called a positive association**. It is also a positive association when a decrease of values of one variable is accompanied by a decrease in values of the other. So, whenever values of the two variables change in the same direction, that is a positive association. **A negative association happens when increase in values of one variable is accompanied by a decrease in values of the other i.e. their corresponding changes are in the opposite directions**. It is important to note that **for an association between variables to have a direction, these variables must be at least on the ordinal level of measurement, i.e. on a measurement level that has directions (lower and higher values). Variables on the nominal level of measurement have no direction of association** because, at this level of measurement, there are no lower and higher values and hence they cannot increase or decrease.

Therefore, associations between variables can take many different forms. but, for the purposes of calculating statistical indicators of these intensities, we can divide them based on whether the direction of their association is constant or changing. With regard to this, associations between variables can be:

- **Monotonic associations** – when the change in values of one variable is accompanied by the changes in values of the other variable, but **always in the same direction**. When values of one variable increase, the values of the other variable either increase or decrease, but this direction remains the same for all values of the first variable. If the values of the second variable increase with increasing values of the first variable, it will remain so for all values of the first variable. If the values of the second variable decrease with increasing values of the first variable, it will also remain so for all values of the first variable. On a graph, the best representation of an association of this type would be a line that never changes direction (both in the horizontal or vertical direction).

• **Non-monotonic associations** – when the change in values of one variable is accompanied by changes in values of the other variable, but these changes have different directions for different values of the first variable. This means that when values of the first variable increase, this will be followed by increasing values of the second variable on some part of the range of the first variable and by decreasing values of the second variable on another part or parts of the range of values of the first variable. In other, words, **the direction of their association changes.** On a graph, the best representation of a non-monotonic relationship would be a line that changes direction either in the horizontal or in the vertical dimension or both.

Another thing to consider when discussing the types of associations between variables, particularly monotonic associations, is the **pace of the change of one variable when the other changes**. This **pace of change can be constant or changing.** A constant pace of change means that whenever the value of a variable changes by a certain amount, there is a corresponding amount of change in the other variable and this correspondence between how much one variable changes vs. how much the other variable changes is constant throughout the range of the variable. If we presented the relationship between values of the two variables on a graph, that relationship would best be described by a straight line. On the other hand, when the pace of change is not constant that means that the size of change of one variable that corresponds to the constant change of the other variable is different for various positions in the value range of the other variables. **Relative to the pace of change, monotonic relationships can be:**

• **linear** – when the **pace of change** of one variable relative to the change of the other variable **is constant.** For example, if changing the value of one variable by 2, leads to a change in the other variable of 6, this correspondence will hold throughout the range of both variables. That means that whatever value we start from, increasing one variable by 2 will be accompanied by the other variable increasing by 6 (or around 6 if the strength of association is less than perfect). On a graph, the best representation of the relationship between two such variables would be a straight line.
• **nonlinear** – when the pace of change of one variable relative to the change of the other variable is changing. For example, with the increase in values of one variable, the other variable can also increase but slowly at first and then faster and faster as the values increase. Or, with the constant rate of change of one variable, the other one can quickly at first and ever slower after that.

In order for a monotonic association between two variables to be linear or nonlinear, the variables whose association is calculated need to be at least on the interval level of measurement. This is because in order to be able to tell whether the magnitude of change of variable values is changing or constant, it needs to be possible to estimate/calculate that magnitude and this requires a unit of measurement of constant size and this only exists starting from the interval level of measurement. **On the ordinal level of measurement, linearity/non-linearity of an association between variables cannot be meaningfully discussed** (Figures 6.1 and 6.2)

One more thing to consider when discussing associations between variables is that **the strength of the association need not be constant throughout the range of the**

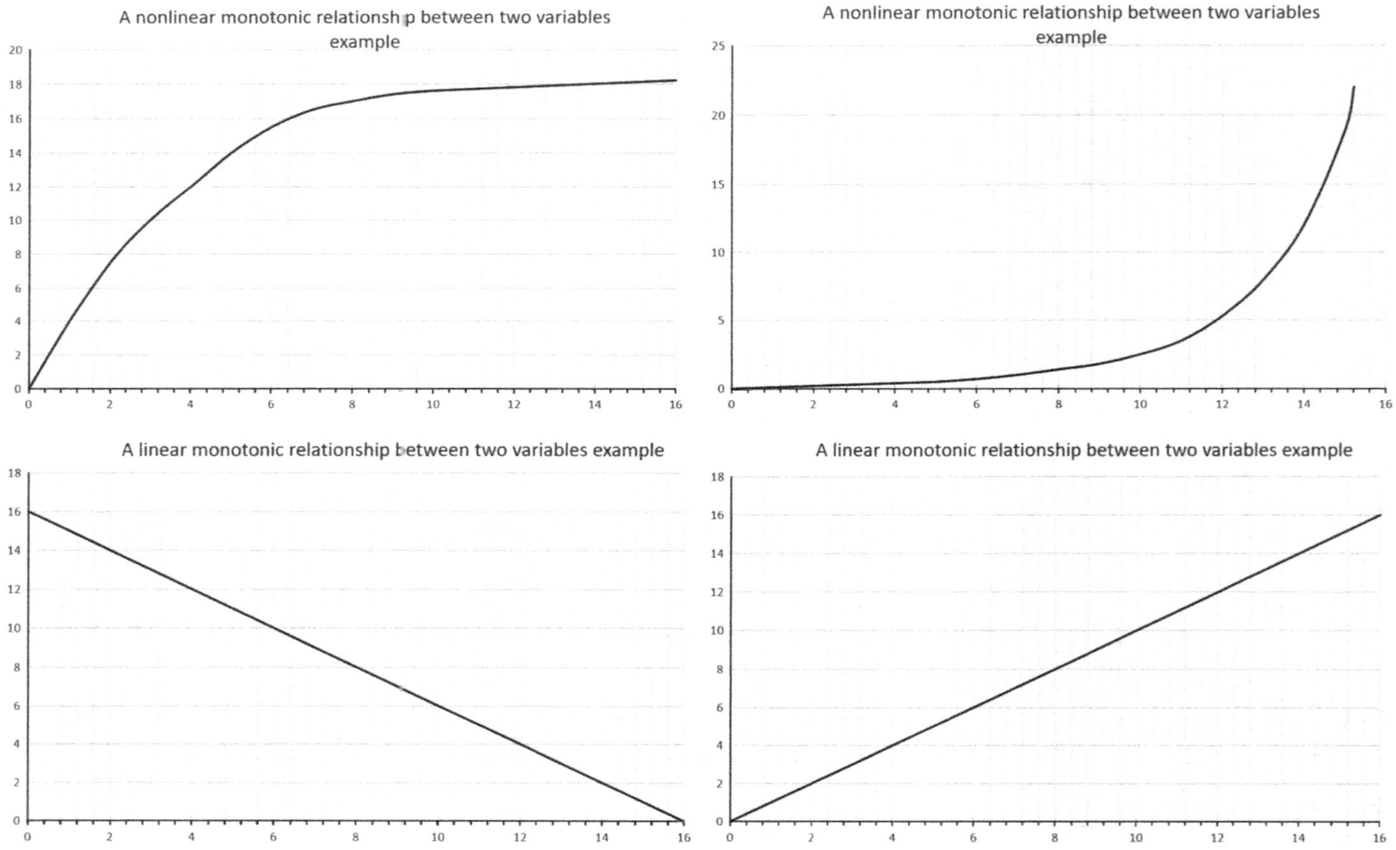

Figure 6.1 Graphical representations of some possible shapes of monotonic relationships. The horizontal axis represents values of one variable, while the vertical represents values of the other variable. Line represents values of the two variables that correspond to each other. For example, from the upper left picture we can see that value 6 from the variable represented by the horizontal axis corresponds to somewhat below 16 on the other variable. The upper two graphs represent nonlinear relationships. We can observe that the line is curved indicating that the pace of change of one variable with the change of the other variable changes. The lower two graphs represent linear relationships. The lower left graph represents a negative association between variables, where the increase of values of one variable is accompanied with a decrease in the values of the other variable. On the other hand, the lower right graph represents a positive association between the two variables, as the increase in values of one is accompanied by an increase in the values of the other. The two nonlinear associations presented in the upper two graphs also represent positive associations between variables. We can see that the increase in values of one variable is also accompanied by increases in the values of the other, however at a changing pace, as the relationships are nonlinear.

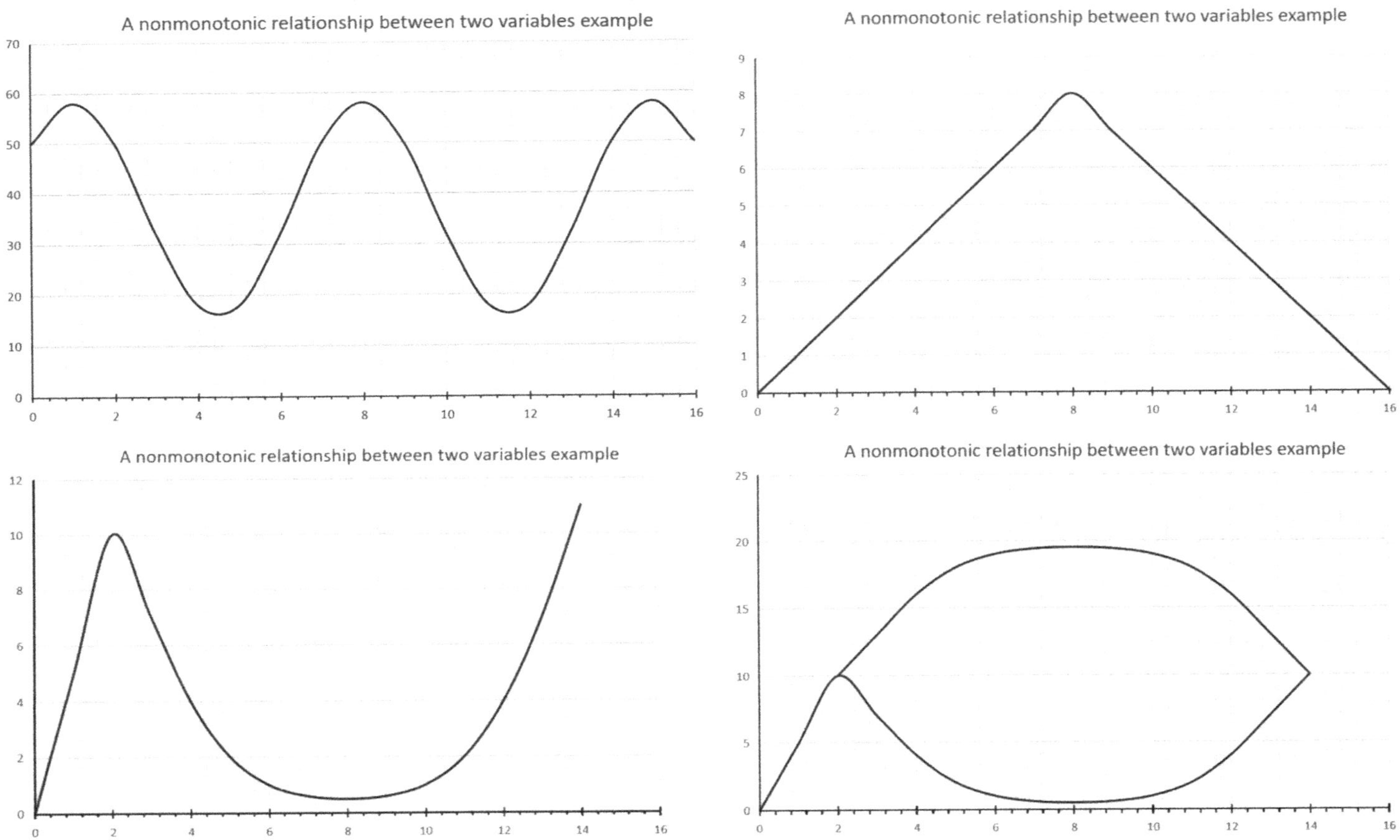

Figure 6.2 Graphical representations of some possible shapes of nonmonotonic relationships. We can see that in these types of relationships at least one variable has two or more corresponding values of the other variables. Due to this, the precision with which values of one variable can be predicted based on the other variable can differ depending on which variable is used to predict which one. For example, if we look at the upper left graph, we can see that if we try to predict values of the variable represented by the vertical axis, based on the values of the variable on the horizontal axis, we can do that perfectly as for each value on the horizontal axis, there is exactly one corresponding value on the vertical axis. On the other hand, most of the values of the variable on the vertical axis have multiple corresponding values on the horizontal axis. For example, if we look at value 40 on the vertical axis, we can see that corresponding values to it on the horizontal axis are 2.3, 6.2, 8.8 and 12.7 (these are just rough readings of the values based on visual inspection of the graph, so they might not be too precise). The situation is similar with the upper right graph (the one with linear parts) and the bottom left graph. On the other hand, with the bottom right graph both variables have more than one corresponding values of the other variable, but not throughout the range. We can see on that graph that values below two on the variable on the horizontal axis have exactly one corresponding value on the vertical axis.

variables in question. While there can indeed be cases when the strength of association between variables (how precisely the changes of one variable accompany the changes of the other) is constant throughout the range of values of variables, there are also cases when the strength of association between two variables differs in different parts of the value ranges of the two variables and even situations where there are value ranges where two variables are associated and ranges where they are not.

While the different types of associations between variables are accommodated for by calculating indicators of association that are appropriate for the type of association, **the way how the changing strength of association between variables is addressed is simply calculating indicators of association for different parts of the variable ranges separately.**

6.3 Scattergram

One very useful tool when analyzing the association between two variables is a scattergram. **A scattergram is a graphical presentation of the relationship between two variables. It is typically a two-dimensional graph where each of the two variables is represented by one dimension.** The **entities are represented by points** whose coordinates on the graph are values on the two variables. In that way, a graph containing a set of dots is produced and from the way dots are grouped we can make inferences about the likely type of relationship between the two variables, the direction of their correlation and also of the approximate intensity of the relationship (Figures 6.3 and 6.4).

Generally, the rules are the following – when correlation is zero, or close to zero, the distribution of entities on the scattergram tends to have a circular shape. The more it deviates from the circle shape, the flatter it gets, the higher the correlation. If the grouping best resembles a straight line, we are likely looking at a linear relationship. If it is a curved line, but one that does not change direction, it is likely a nonlinear monotonic association. Finally, a distribution of dots resembling a curved line that changes direction indicates a non-monotonic association between variables.

6.4 Correlation coefficient

Statistics used as indicators of strength (or intensity) of association between two variables are called correlation coefficients. **A correlation coefficient can typically have values between 0 and 1 and can be positive or negative i.e. + or -. The value of the correlation coefficient and its sign are interpreted independently of each other. A correlation coefficient typically indicates two things:**

- **The intensity of the correlation** i.e. the strength of the association between two variables is indicated by the (absolute) value of the coefficient. This value can range between 0 and 1. A correlation of 0 indicates that there is no association between variables, while a correlation of 1 indicates that there is a total or a maximum association between variables. In other words, when correlation is 0, this means that predicting the values of one variable based on the other is no better than chance i.e. no better than assigning random values to the variable we are predicting. On the other hand, a correlation of 1 means that we can predict values of one variable based on the other perfectly, without any errors. Correlations between 0 and 1 mean that we will make certain errors in predicting values of one variable based on values of

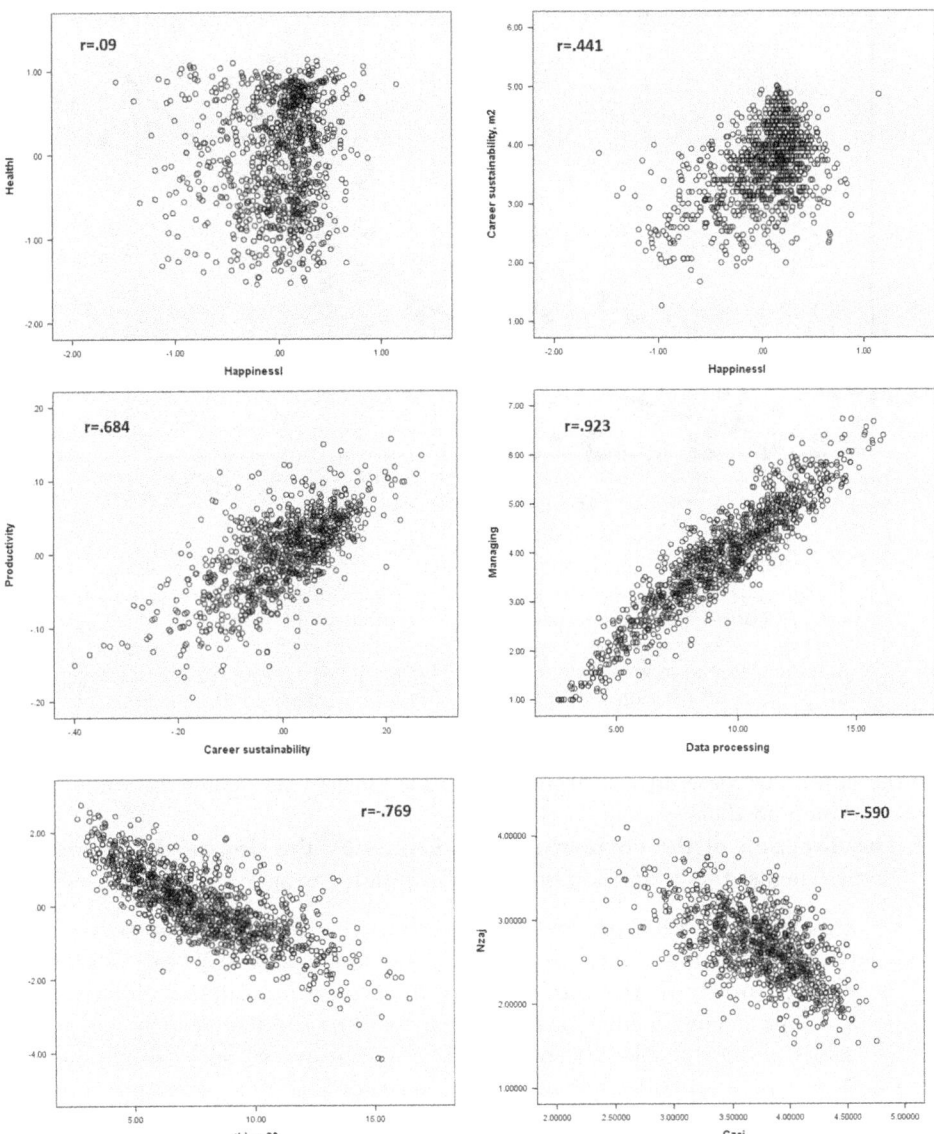

Figure 6.3 Examples of scattergrams depicting correlations between two variables of various intensity and directions. We can see that as the correlation between variables becomes stronger, so does the general shape of the dots on the scattergram become flatter and, in the opposite direction, as the correlation becomes weaker, so does the shape of the configuration of dots on the scattergram become rounder. The upper 4 scattergrams depict positive correlations, while the bottom two depict negative correlations. We can see that, when the correlations are negative, high values of one variable, correspond to low levels of the other variable. The data comes from various studies conducted by the authors.

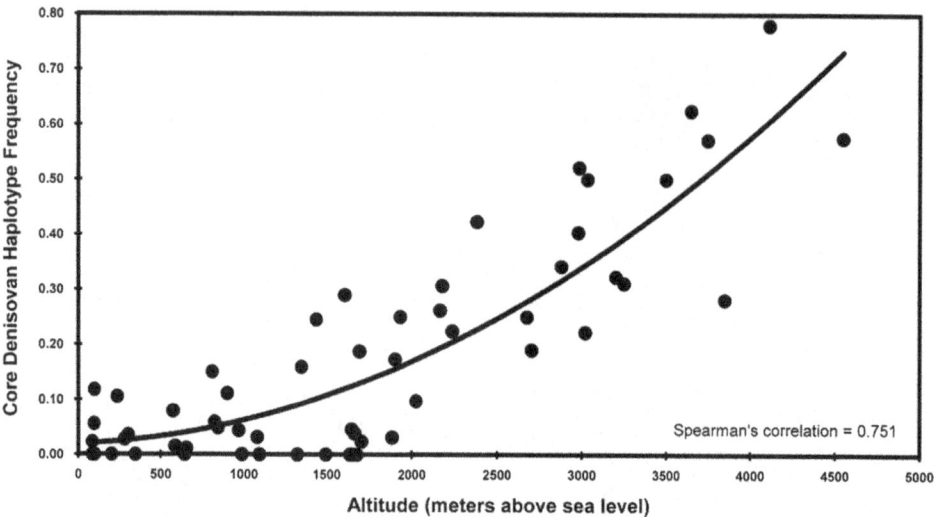

Figure 6.4 An example of a scattergram depicting a slightly nonlinear, monotonic, positive correlation. The image is reprinted from Hackinger et al. (2016), with the permission of authors and it is made available under the Creative Commons Attribution 4.0 International License (http://creativecommons.org/licenses/by/4.0/). Aside from removing the original figure number, to prevent confusion with the numbering of figures in this book and graphical formatting for the purposes of inclusion in this book, no changes were made to the content of the figure.

the other, but these errors will be smaller and smaller the higher the value of the correlation coefficient.

- **The direction of the correlation is indicated by the sign of the coefficient** i.e. by whether the correlation coefficient is positive or negative:

  - A **positive correlation** coefficient indicates that the correlation between the two variables is positive. A positive correlation means that **variable values change jointly in the same direction.** Increase in values of one variable is accompanied by the increase in values of the other and vice versa – a decrease in values of one variable is accompanied by a decrease in values of the other.
  - A **negative correlation** coefficient indicates that the correlation between variables is negative. A negative correlation means that **variable values change jointly in opposite directions.** An increase of values of one variable is accompanied by a decrease in values of the other.

Finally, **a zero (0) correlation indicates that there is no correlation between variables**. Changes in the values of one variable are not accompanied by any specific or predictable changes in the other variable i.e. **variables change their values independent of each other**. However, it should be noted here that there exist different correlation coefficients (to be discussed in the further part of the book) designed for describing intensities of different types of associations between variables. Due to this, **a zero correlation might also mean that the type of association between variables is not the one that can be detected/validly indicated by the correlation coefficient that**

was calculated. A very common situation of this type is the one where the association between variables is non-monotonic, but we calculate a correlation coefficient designed for describing linear associations between variables (linear correlations). In such a case, a linear correlation coefficient would likely indicate a zero correlation, when in reality there is a non-monotonic correlation of some intensity.

It should also be noted that **not all correlation coefficients have a direction and therefore a sign**:

• **Correlation coefficients used to express the intensity of a non-monotonic association between variables do not have a sign** i.e. a direction, because in a non-monotonic relationship, the direction of association is not constant, but changes. Therefore, it is positive on a part of the variable value range and negative on the other and also possibly zero on another part of the variable value range, so a single correlation direction indicator would not do.

• **Correlation coefficients used to express the intensity of association between nominal variables also do not have a direction or sign**. As noted earlier, a direction of association needs data in which it can be established which values are lower and which are higher (thus making increases and decreases in values possible). This is not the case with nominal variables, in which for any two values we can only establish whether they are equal or not and cannot compare their sizes meaningfully. Hence, correlation coefficients cannot be positive or negative with these variables.

One issue to note, with indicators of the direction of the correlation is that in the current protocols for writing numbers, a positive sign is not written in front of a positive number i.e. when we want to write, for example, +3, we only write 3. The sign of the correlation coefficient is only written for negative values. Due to this, **when interpreting statistical results presented in literature, it may require additional attention to details of the presented results to differentiate between positive correlation coefficients and correlation coefficients without a sign i.e. that do not have a direction**. On the other hand, when we see a negative correlation coefficient, it is definitely a coefficient with a direction.

When interpreting the intensity of a correlation coefficient, there are no set or natural rules about what exactly constitutes a small correlation and what is a large correlation coefficient, however there are various recommendations on how to interpret them. Probably the one set of recommendations that is most commonly found in literature is the one proposed by Cohen (1988). **Cohen's widely accepted recommendation is that correlation coefficients of .1 be interpreted as weak/small, a coefficient of .3 should be considered moderate and a correlation of .5 should be considered large.** If we convert these recommendations into intervals, we could say that correlations up to .2 can be considered weak/small, those between .2 and .4 can be considered moderate and those above .5 are large. However, what is large and what can be considered small can largely depend on the point of reference. In his review, **Taylor (1990)** mentions studies that reported correlations above .9 in the medical field and lists recommendations for interpretation of correlation coefficient sizes according to which a **correlation size below .35 would be considered weak, those between .36 and .67 modest or moderate and those above .68 would be considered high, with correlations above .90 being considered very high.** On the other hand, a review of correlations obtained in the field of psychology published in a commentary by Hemphill

(2003), considers results of a large number of meta-analytic studies on research in **psychology** that categorized correlations obtained in these studies into three equally large categories according to sample size and reported **that around 1/3 of the studies reported correlation coefficients up to around .2, another 1/3 between around .2 to .3 and the 1/3 of studies with the highest correlations reported correlations between .3 or .35, up to .6 or .78 depending on the groups of studies reviewed.** Therefore, **practical guidelines for interpreting correlation coefficients can differ somewhat depending on the area one is working in.** Having this in mind, in practice it might be the wisest to use the interpretation rules that are accepted in the specific area one is working in.

When we wish **to estimate the likely size of the correlation coefficient in the population**, it is most commonly done by **testing the statistical significance of the null hypothesis that the correlation in the population equals 0 (zero)** i.e. that the sample was taken from a population in which the correlation is zero. Under the central limit theorem approach, the testing is done by first calculating the standard error of the correlation coefficient and then creating a confidence interval encompassing the desired portion of the theoretical sampling distribution according to this theorem. The most commonly used **formula for calculating the standard error of the correlation coefficient** (of the Pearson product-moment correlation coefficient, to be described in the next chapter) is:

$$SE_r = \sqrt{\frac{1 - r^2}{N - 2}}$$

where SE_r is the standard error of the correlation coefficient, r is the size of the correlation coefficient and N is the sample size i.e. the number of entities in the sample. While this is a general formula, **when calculating the standard error of the correlation coefficient for the purpose of testing a null hypothesis that the correlation coefficient equals zero, the r in the equation is 0** (because we are creating a confidence interval for a correlation of zero). Also, for practical purposes and when we are dealing with large sample sizes typically found in studies in social sciences, we can also, for the sake of simplicity, remove the –2 from the equation, because dividing something with 202 and with 200 (or dividing with 1002 vs. 1000) does not produce any appreciable difference, **this equation can be simplified to**:

$$SE_r = \frac{1}{\sqrt{N}}$$

We then multiply this standard error with an appropriate coefficient depending on the critical level of statistical significance we wish to use (1,96 for the .05 level and 2,56 for the .01 level) and thus obtain the minimum size the correlation coefficient needs to have to pass the statistical significance threshold.

While using this equation can be useful for making quick back-of-the-envelope estimation of how large correlation coefficients need to be to pass the statistical significance threshold on a sample of a certain size, **research studies usually either clearly mark correlation coefficients that pass the significance threshold i.e. that are statistically significant,** as this is usually called in the scientific jargon or report specific statistical significance levels for each correlation coefficient. In the latter case we can decide whether a correlation coefficient is statistically significant or not i.e. whether we should accept or reject the null hypothesis (which is usually that the correlation in the population is 0) based on whether the statistical significance level (as explained in the previous part of this book) of the correlation coefficient is above or below our statistical significance threshold. For example, if the statistical significance of a correlation coefficient is .023 and our statistical significance threshold is .05, we will reject the null hypothesis. If, on the other hand, the statistical significance of the correlation coefficient is .22, we will accept the null hypothesis. **If we decide to accept the null hypothesis, we are than obliged to treat the correlation as if it is zero or more precisely, we need to accept that the data at hand does not allow us to conclude that there is a non-zero correlation in the population. If we decide to reject the null hypothesis, we then treat the population correlation level as being approximately equal to the value of the correlation coefficient that we obtained on the sample.** Also, for additional precision, a confidence interval can be created around our sample correlation coefficient in the way that was described in the previous part of this book.

Another way to test the null hypothesis about the population size of the correlation coefficient is through the **bootstrapping procedure**, as described in the chapter on statistical significance. In this case, **a bootstrap confidence interval of desired size (most commonly 95% or 99%) is created** by drawing a predefined large number of samples with replacement from the study sample and establishing a sampling distribution of the correlation coefficient based on that. **If the value specified by the null hypothesis (i.e. zero) lies within the confidence interval created in this way, we accept the null hypothesis and if it lies outside the confidence interval, we reject the null hypothesis.**

6.5 Types of correlation coefficients

It was said earlier that a correlation coefficient is not a single statistic type, but rather that there are many types of correlation coefficients. This is both due to different authors proposing different ways of calculating the correlation between variables and the fact that there are different types of association between variables (as was discussed previously) and these different types of association cannot be encompassed by a single formula at this point in the development of the field of statistics. Some of the most commonly used types of correlation coefficients will be presented in this book, but the reader should be aware that there are many, many more correlation coefficients proposed and being proposed by various authors, so this should in no way be treated as an exhaustive list of measures of association between variables.

Pearson's product-moment correlation coefficient or Pearson correlation coefficient, for short, is probably the most well-known and the most widely used correlation coefficient. It is based on the assumption that the **relationship between the two variables it is used on is linear** and it expresses the strength of the linear association between two variables. In literature, we can find it used both as a stand-alone indicator of linear correlation between variables and as a part of many more complex statistical procedures. It is **typically denoted as "r".** It is **a parametric statistical measure** requiring that the variables be at least on the interval level of measurement and that the data be normally distributed. It is calculated as the sum of products of z scores of entities on the two variables divided by the number of degrees of freedom (or, if we want to keep thing simple, – the number of entities, the difference is just 1 and thus practically completely negligible for large samples):

$$r = \frac{\sum_{i=1}^{N} z1_i * z2_i}{N - 1}$$

In this equation r is the Pearson's product-moment correlation coefficient, z1 and z2 are values of variables whose correlation we are calculating expressed as z scores and N is the number of entities in the sample i.e. the number of pairs of data in z1 and z2. It should be noted that, while we are presenting this formula of the Pearson correlation coefficient here for the sake of simplicity, versions of the formula exist (and can be derived from this formula!) where data is entered into the equation in its raw form without first converting it into z scores.

As is generally the case with correlation coefficients, **Pearson's correlation ranges from 0 to ±1 (+1 or −1)**. If we look at the formula for this correlation coefficient, we can observe a couple of things:

- Since the coefficient is based on the product of z scores, this means that **measurement units of the two variables or scales they are on do not matter**. In the process of calculating the correlation coefficient (or before calculating it, if we use the formula presented earlier), variable values are converted to z scores and that means that into the calculation of the coefficient they enter with identical standard deviations (i.e. variances) and identical means. This means that Pearson correlation coefficient can be calculated for any two variables as long as their distributions are normal (or sufficiently close to normal.

- Since it is a product of the z values of entities on the two variables, when both z values are positive and when both z values are negative their product will have a positive value. However, when one of the z scores is positive and the other is negative, the product will have a negative value. So, if the sum of products of z values of entities that have values in the same half of the distribution (upper or lower) exceeds the sum of products of z values of entities that have values in opposite halves of distributions of the two variables, the correlation coefficient will be positive. If the later exceeds the former, the correlation coefficient will be negative. However, if these two sums are equal, the correlation coefficient will be 0. This means that **even in situations when the correlation between two variables is positive, there can be aberrant cases that have high values on one variable and low on the other.** And similarly, **even when correlation coefficient is negative, there can be cases that have values with similar positions on distributions of both variables** (high values on both or low values on both). However, such cases will reduce the correlation coefficient, but the lower the correlation coefficient the more opportunity for such cases to exist.

- The previous also means that if our sample actually comes from two different populations, one on which there is a positive correlation and the other one with a negative correlation between variables, on the sample created from those two groups, we will have a zero correlation (or a much lower correlation in one or the other direction, depending on how high correlations within groups are and on what are the relative group sizes).

- Products of numbers above 1 are larger than either of these numbers, while products of numbers below 1 are lower than either of those numbers. Since the dividend in the equation for the Pearson correlation coefficient is the sum of these products, this means that products of larger values will contribute to the total sum much more than products of smaller values. That is why **entities with extreme values on both variables can have a very substantial and disproportionate influence on the value of the correlation coefficient**. One extreme case with the relationship of values in one direction can nullify the effects of multiple cases with the relationship of values going in the other direction. For example, if we had a single case with the z scores of +7 on both variables, that would make the product 49. To bring the total sum to zero i.e. to nullify this value, we would need 196 cases that have z score value of .5 on one of the variables and -.5 on the other. Of course, if such was the total sample, we would likely not be calculating Pearson correlation at all, because the requirement that the distribution be normal is not fulfilled, even if these were just a part of the total sample. However, we use this example to show how high z score values can disproportionately affect the value of the Pearson correlation. This is **particularly important to note because extreme values in research are quite often the product of data entry mistakes or erroneous measurements and these should be resolved prior to calculating correlations**, even if they are not detected when inspecting the shape of the distribution.

Square of the Pearson correlation coefficient is called the coefficient of determination and it **is interpreted as proportion of variance the two variables share (under the assumption that their association is completely linear).** We should remember that correlation implies that two variables change together i.e. that changes in one variable are accompanied by changes in the other, but that this is a matter of degree. When this degree is presented as a proportion of the total variability expressed as variance, the part of it that happens jointly in the two variables is expressed by the coefficient of determination. Since the variances of the two variables for which the Pearson correlation is calculated are equalized prior to or in the process of calculating the correlation through conversion to z scores, the proportion of the variance one variable shares with the other is equal to the proportion the other shares with the first variable i.e. this proportion is equal for both variables. The coefficient of determination is often used as an indicator of the magnitude of association between two variables in the scope of various multivariate statistical procedures (not covered in this book), but is also sometimes used as a standalone indicator of the intensity of association between variables, as it is preferred by many researchers, because it is a proportion and thus sometimes more conveniently interpretable than the correlation coefficient.

Pearson's correlation coefficient is also calculated in the scope of various multivariate statistical procedures where it is used as an indicator of correlation between variables derived from the study variables. **It is then given different names, depending on the nature of the variables for which the correlation is calculated and how they are derived.** These names include "canonical correlation", "coefficient of multiple correlation", "factor loading", "structure coefficient", "cross-structure coefficient", "partial correlation", "semi-partial correlation" etc.

Spearman's rank correlation coefficient is a coefficient of **correlation intended for variables on at least the ordinal level of measurement** and based on the assumption that **the association between the two variables is monotonic.** It is **a**

nonparametric statistic, meaning that it makes no assumptions about the shape of the distribution of variables between which the correlation is calculated. There are different formulas for calculating this coefficient, however probably the simplest way is to rank the data i.e. convert the data into ranks on each variable separately and then calculate the Pearson correlation coefficient between these ranks (in case the formula for the Pearson correlation coefficient presented in this book is used, the ranks need to be converted to z scores first, using the regular formula for converting raw values to z scores). Ranks are created by simply assigning the value 1 to the entity with the smallest value, then rank 2 to the second smallest and so on, up to the largest value. This is done for each of the two variables separately. Spearmen's rank correlation coefficient is **typically denoted with the Greek letter rho (ρ),** though readers should take care when coming across rho in the literature as **the same symbol is used to denote the population value of the Pearson's correlation coefficient** (in contrast to the sample value which is then denoted with r) and possibly other statistics.

Point-biserial correlation coefficient is the **Pearson product-moment correlation coefficient calculated between one binary and one interval variable.** It is **typically used to show the degree of overlap between distributions of two groups and, at the same time, the size of the difference between the arithmetic means of those two groups.** That is why it is often referred to in literature simply as the **effect size** (when the idea behind calculating it is estimating the size of the difference between means of two groups). In this situation, the binary variable contains information on the group membership of each entity i.e. one group is denoted with one value of the binary variable and the other with the other value of the binary variable. Calculating the point-biserial correlation coefficient then implies that one group will be marked with 0 and the other group with 1 (or actually with any two numbers) and given that binary variables can be treated as data on any level of measurement, both the interval variable (the one on which the entities were measured or assessed) and the binary variable can be entered into the calculation of the Pearson correlation, only the resulting statistic will not be called Pearson correlation coefficient but point-biserial coefficient.

As is the case with the Pearson correlation coefficient, **the point-biserial correlation also ranges from −1 to +1, but the interpretation is a bit different. A positive and a negative correlation have little meaning when one of the variables is binary and whether the coefficient itself will be positive or negative depends solely on which of the two groups was assigned a larger number. If the group that was marked with the higher number has the higher mean, than the point-biserial correlation coefficient will be positive. If the group that was marked with the lower number has the higher mean, the point-biserial correlation coefficient will be negative.** If we reverse the assignments of numbers to groups (we mark the group that was previously assigned the higher number with the lower and vice versa), the sign of the point-biserial correlation coefficient will change. For example, if we had a group of people from London and a group of people from New York and we wanted to compare their mean weights we could mark Londoners with 0 and New Yorkers with 1, and if it turned out that New Yorkers had higher mean weights, then the point-biserial correlation coefficient between the city of residence and weight would be positive. If it were Londoners who happened to have a higher mean weight, the point-biserial correlation coefficient between the city of residence and weight would be negative. This is because the Londoners, who have the higher mean are marked with 0, which is the lower number of the two used to designate the two groups. In a similar fashion, if we obtained a point-biserial correlation of, for example, -.75 from a study comparing means of two groups on some interval variable, this would mean that the group marked with the lower number has the higher mean of the two.

When considering the intensity of the point-biserial correlation coefficient, it was noted earlier that it represents the level of overlap between distributions of two groups on an interval variable (with normal distributions, it should be noted, as this is a parametric statistical procedure!). **When the point-biserial correlation coefficient is 0, that means that distributions of the two groups are completely overlapping** i.e. **that there is no difference either between means of the two groups or between their distributions.** They completely overlap. **When the point-biserial correlation coefficient is 1, that means that there is no overlap between distributions.** In other words, if we know the values of an entity on the interval variable, we can perfectly, without any errors, predict group membership of that entity on the binary variable i.e. which of the two groups it belongs to. The reverse does not apply, as entities in the same group can have a range of different values on the interval variable. When the point-biserial correlation is 1 (or -1) we can only predict a range of values within which the value of the entity will lie if we know its group membership i.e. its value on the binary variable. Of course, when the point-biserial correlation is between 0 and 1 (or between 0 and -1, which is the same!) we will be able to predict group membership based on the values of the interval variable better than we would do by random guessing, but the lower the correlation, the more errors we will make. This is because the overlap between distributions of the two groups on the interval variable will increase as the point-biserial correlation decreases. (Figure 6.5).

One **requirement** for the calculation to be done in this way is that **the binary variable be truly binar**y i.e. that the two values of the binary variable are not merely broad categories for an underlying continuum of different values. For example, if we wanted to compare residents of London (group 1) and residents of New York (group 2) with regard to their body weight we could collect a sample of Londoners and a sample of New Yorkers, measure their weight and then calculate the point-biserial correlation coefficient between the city of residence, i.e. the binary variable containing information whether the person is a Londoner or a New Yorker, and weight, which is a ratio variable. This would be adequate because the city of residence, with possible values London and New York is a true binary variable. On the other hand, if we, for example, wanted to explore whether students who have failed a written exam in some school subject and those who have passed it have different attitudes towards that subject (let us say that attitude was measured using a valid test of attitude towards that subject and that it is a normally distributed interval level measure) and to do this, we created a binary variable containing information on whether a person has passed or failed the course derived from his/her score on the written exam, it would not be appropriate to use point-biserial correlation coefficient. This is because the binary variable containing information on whether a person has passed or failed the test is not a true binary variable i.e. a true dichotomy, but is derived from the underlying continuum of different test scores. People who have failed have achieved a whole range of different scores, which only have in common that they are all below the threshold needed to pass the test. And it is a similar situation with those who have passed the test – they also have a range of different scores (meaning also range of different levels of knowledge of the course) but it is only that all these scores are above the threshold needed to pass. In such situation, the correlation coefficient that should be calculated is **the biserial correlation coefficient.**

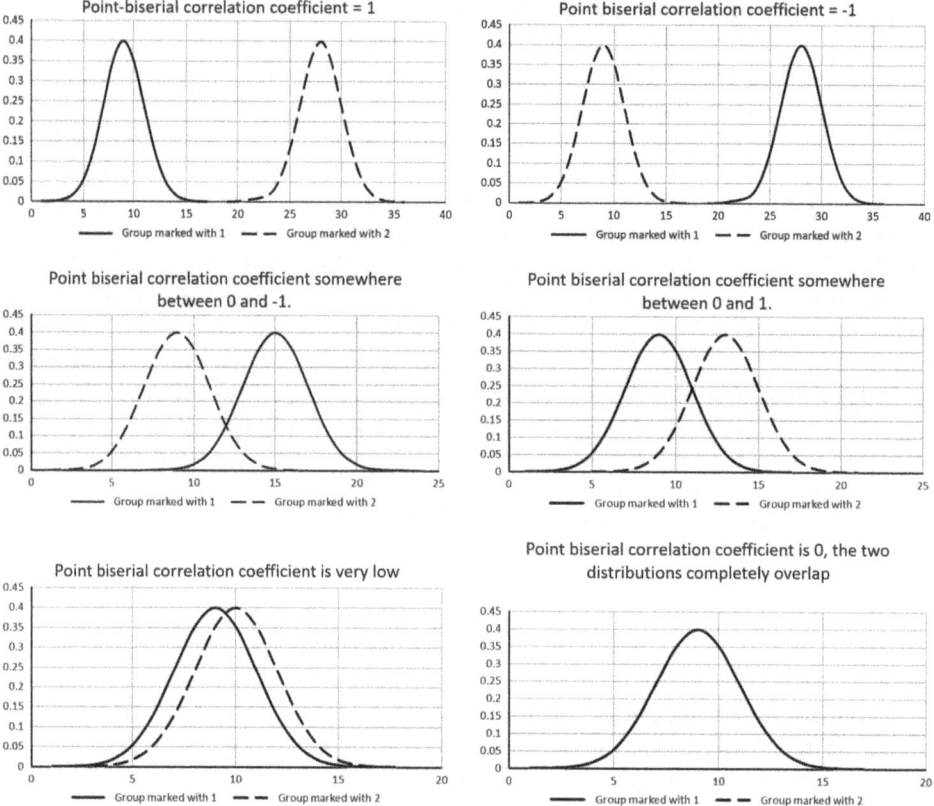

Figure 6.5 A graphical representation of various degrees of overlap between distributions of two groups and their corresponding point-biserial correlation coefficients. We can see from the picture that when the two distributions do not overlap, the correlation is 1 or -1. It is 1 when the group marked with a higher number has the higher mean. It is -1 when the group with a higher number has the lower mean. When the two distributions overlap, but not completely, correlation coefficient takes values between 0 and 1 (or 0 and -1, depending on which group is marked with a higher number for the calculation of the coefficient). Finally, when two groups overlap, the coefficient is 0.

Although named similarly, the formula for calculating the biserial correlation coefficient is quite different from the one for calculating the point-biserial coefficient. That said, in situations such as the one in the example, a much better option is to simply calculate the Pearson or Spearman correlation coefficient between the original two interval variables and not bother with creating an artificial dichotomy at all.

Eta coefficient (η) is a coefficient **that can either be used as a measure of association between one nominal and one interval variable or of monotonic non-linear or non-monotonic association between two interval variables. The nominal variable in the eta calculation may be truly nominal, but it can also be a discrete variable with not too many categories** (for example, an interval variable reduced into several groups). **When the relationship between two interval variables is linear, eta and the Pearson correlation coefficient are equal,** however, when it is not linear, eta will be higher than the Pearson correlation coefficient. Also**, when the relationship between the two variables is non-monotonic two separate values of eta are calculated** – one for predicting the values of the first variable based on the second and one for predicting the values of the second variable based on the first. This is because with non-monotonic relationships, due to the changing direction of the association between variables, the precision with which one variable can be predicted based on the values of the other, is often different from the precision that can be achieved when that other variable is predicted based on the first. For example, in the previously used example of the relationship proposed by the Dunning-Kruger effect (Kruger & Dunning, 1999) between the self-assessment of competence for something and real competence, we would likely be much better at predicting how a person will self-assess his/her competence when we know the real competence of the person than we would be able to predict how competent a person is based on that person's self-assessment of his/her own competence. This is because as real competence increases, self-assessment of competence first increases somewhat, than tends to decrease and then increases again with further increase of competence (this refers to the changes in perceived test score as a function of actual test score as reported in studies 2 and 3 of this paper). In such a situation, high self-assessments correspond both to high and to low levels of competence, but such double correspondence does not exist when predicting self-assessment based on real competence. When **calculating eta, the variable that is used for predicting the other variable is called the independent variable, while the variable being predicted is called the dependent variable**. In the two calculations of eta, therefore, variables switch places with each being once a dependent and once an independent variable. However, **when eta is used to calculate correlation between a nominal and an interval variable, typically only the eta coefficient obtained with the nominal variable as independent and the interval variable as dependent is reported (or calculated).** This property, that the statistic will have different values depending on which variable is treated as dependent and which as independent is called **asymmetry** (in contrast to statistics like the Pearson correlation coefficient which are symmetric). However, although naming of variables as independent and dependent is the same as variables are named in experiments, we should remember that correlations alone cannot establish causal relationships between variables and this is also the case with eta.

When **used to express the level of association between a nominal and an interval variable, eta can be interpreted as an indicator of the level of overlap between distributions of the groups of entities** with different values of the nominal variable on the interval variable or equally as **an indicator of the size of the differences between arithmetic means of these groups**. Each group consists of entities with the same value on the nominal variable, so there are as many groups as the nominal variable has values (one value of the nominal variable forms one group). **When eta is 0, arithmetic means of groups all equal the mean of the whole sample** (all groups together) **and their distributions overlap**. As eta becomes higher, so does the level of overlap between distributions (on the interval variable) of groups decrease.

Eta is a parametric statistic meaning that interval variables are expected to be normally distributed. When eta is used to calculate the level of association between a nominal and an interval variable, **there should not be values of the nominal variables with too low frequencies**. What exactly is too low generally depends on the context of the specific study area, but there should generally be no groups with numbers of entities in the single digits. Because eta is a coefficient of non-monotonic association (and also because one of the variables is nominal)**, eta coefficient does not have a sign** i.e. cannot be positive or negative.

When **used to express the size of differences between means of multiple groups, eta squared is often used along with or instead of eta**. Eta squared is obtained by squaring the eta coefficient (multiplying eta coefficient with itself). It is interpreted in a similar fashion as the square of the Pearson correlation coefficient – the coefficient of determination, as the proportion of variance shared by the two variables.

Phi correlation coefficient is the **Pearson correlation coefficient calculated between two binary variables.** With this correlation coefficient, **the sign indicates which category of one variable tends to be associated with which category of the other variable**. Given that both variables are binary and that their categories are marked by numbers, with each variable, one category will be marked with a higher and one with a lower number. If entities belonging to the category marked with the higher number on one variable tend to also belong to the category marked with a higher number on the other variable, then the sign of the phi coefficient will be positive. In the reverse situations – when entities marked with a higher number on one variable tend to belong to the category marked with the lower number on the other, then phi coefficient will be negative. Similar to how it is interpreted in the case of the point biserial correlation coefficient, **the sign of the coefficient should not be interpreted as indicating a positive or negative correlation, but only as an indication of how the categories of the two binary variables are associated**. The way numbers are assigned to the two categories of a binary variable i.e. which category receives the lower and which receives a higher number is most often arbitrary meaning that the category marked with the lower number could have equally well been marked with a higher number and vice versa. Due to this, it does not really make sense to interpret the sign of the phi correlation coefficient as the direction of the correlation as it can change simply by marking the categories with numbers differently. For example, if we had two binary variables - one containing the information on whether a child is a primary or a secondary school student, let us call it "school" and another variable containing information on whether the child prefers blue or white t-shirts, let us call it "t-shirt", it would not make any sense to say that the correlation between school and t-shirt is positive or that it is negative. It would however be meaningful to say that for example, primary school students tend to prefer white t-shirts, while the secondary school students tended to prefer blue ones more, of course if we obtained a correlation coefficient indicating this. In the same manner, if our two variables were whether a person belonged to group A or B in a school ("group") and whether he/she has passed or failed a test ("test"), it would not be very informative to say that there is a positive or a negative correlation between variable groups and the test. However, it would be meaningful to say that more people from one group passed the test then from the other (if, of course, that were the case).

Contingency coefficient is a **correlation coefficient intended for describing the level of association between two nominal variables.** In a way, it can be considered to be an adjustment of the phi coefficient for (nominal) variables with more than two categories. However, unlike the phi coefficient, it does not have a sign i.e. it is always positive. In general, a 0 contingency coefficient indicates that two nominal variables are not associated i.e. belonging to a certain category on one variable does not indicate a higher or lower chance to belong to any particular category on the other variable. On the other hand, the larger the correlation coefficient, the larger the association between the two variables meaning that entities belonging to a certain category on one variable will be more likely to belong to a certain category (or categories) of the other variable. Unlike the situation with the phi coefficient, **with the contingency coefficient, the information about which category of one variable is associated with which category of the other variable cannot be deciphered from the coefficient alone, but typically requires the inspection of the cross tabulated frequencies of categories of the two variables.** It is also important to know that the **contingency coefficient will never have the value of one, even if there is perfect association between the two variables, but will only approach it.** Also, the **size of the contingency coefficient depends on the number of categories of the two nominal variables – the contingency coefficient tends to get higher with the higher number of categories of the two variables and also tends to be smaller with nominal variables with a smaller number of categories.** It can often be heard that researchers recommend using the contingency coefficient **only in situations when nominal variables have at least 5 categories each.**

Cramer's V is another popular correlation coefficient for **describing the level of association between two nominal variables.** Cramer's V is calculated using a formula that allows for its upper limit to be 1. This means that **in the case of a perfect association between variables, the value of Cramer's V will be 1** (unlike the contingency coefficient that cannot reach 1) and this is why it is preferred by many researchers as a measure of association between two nominal variables.

6.6 Let us apply what we learned so far!

Let us now try to apply what we have covered in this chapter through a couple of exercises. Please refer to the start of the book for the general instruction for completing the exercises. Our suggestion is that you first read the excerpt and the statements and provide your own answer. You may write it in the Answer column, and after that look up the answers and compare your own answers with them.

Exercise L. Correlations. (Andelković et al., 2012)

Variables		*Trust*	*Safety*	*Independence/autonomy*	*Companionship*
Quality of Life	Social relations	.209*	.225*	.126*	.205*
	Physical health	.063*	-.086*	-.052*	-.052*
	Living conditions	.034	.056*	.096*	.124*
	Mental Health	.062*	.073*	.077*	.099*

* – statistically significant at least at the .05 level.

Table is based on authors' own data from the study published in (Andelković et al., 2012).

L	Statement:	*Answer*
L1.	As Trust increases, Living conditions also tend to increase (improve).	
L2.	As Independence/autonomy increases, Physical health tends to decrease.	

(Continued)

L	Statement:	Answer
L3.	As Physical health decreases, Companionship also tends to decrease.	
L4.	The lower values on Mental health tend to be associated with lower values on Trust.	
L5.	Safety is on the nominal level of measurement.	
L6.	All presented correlations are statistically significant.	
L7.	The intensity of all the presented statistically significant correlations is low, if we asses them according to Cohen's recommendations.	
L8.	Trust is not associated with Living conditions.	
L9.	Correlations of variables presented in columns tend to be more intensive with Living conditions than with any of the other variables presented in rows.	
L10.	Lower values on Physcial health tend to be associated with higher values on Safety.	

Exercise M. Correlations. (Hedrih, 2011)

	R	I	A	S	E	C
R	–					
I	.45	–				
A	.26	.60	–			
S	.14	.40	.48	–		
E	.37	.25	.26	.62	–	
C	.64	.30	.14	.20	.54	–

All correlations higher than .07 are statistically significant at least at the .05 level. The correlations are Pearson's product-moment correlation coefficients.

Table presents the correlations between Holland's vocational interest types. Table based on data from the study of (Hedrih, 2011).

M	Statement:	Answer
M1.	R and I are in a correlation of moderate intensity, judging by Taylor's recommendation.	
M2.	If correlation coefficients were displayed along the main diagonal, instead of dashes, all these correlations would be 0.	
M3.	All presented correlations are statistically significant.	
M4.	Correlation between A and C is weak, judging by Cohen's recommendation.	
M5.	A and S are not associated.	
M6.	E is the cause of C.	
M7.	When R increases, C increases also.	
M8.	When C decreases, I also decreases.	
M9.	If correlations were presented in the upper triangle of the table, they would be the same as correlations presented in the lower triangle.	
M10.	The presented correlation coefficients are Spearman's rank correlation coefficients.	

Exercise N. Correlations. (Radanović et al., 2021)

Descriptive statistics and intercorrelations of variables in the study

	M	SD	1	2	3	4	5	6	7	8	9	10	11
Children													
1. CovFear_C	2.83	0.83											
2. Age_Children	12.78	3.57	-.15**										
Parents													
3. Distress NSE	2.76	1.00	.03	.04									
4. Distress CF	3.02	0.67	-.03	.05	.06								
5. Distress PD	2.68	0.92	.37**	.02	.21**	.07							
6. SelfEf	5.62	0.99	.10*	-.14**	-.18**	.11*	-.15**						
7. ER_CR	5.19	1.25	.10	-.07	-.19**	-.02	-.06	.32**					
8. ER_ES	3.36	1.40	.10*	-.04	-.18**	.03	.09	.00	.30**				
9. SomAnx	1.39	0.57	.19**	.08	.17**	.04	.40**	-.21**	-.09	.10			
10. CogAnx	1.70	0.63	.17**	.06	.21**	.07	.49**	-.22**	-.13*	.07	.70**		
11. CovFear_P	2.73	0.74	.49**	.03	.03	.08	.66**	-.07	-.02	.07	.37**	.44**	
12. PanParent	3.98	0.63	.24**	-.27**	-.04	-.02	.08	.24**	.34**	.03	.03	.04	.12*

$*p < .05$; $**p < .01$.

Note. CovFear_C = Children's Fear of COVID-19; Age_Children = Children's age; Distress NSE = Parental distress due to the national state of emergency; Distress CF = Parental distress due to the curfew; Distress PD = Parental distress due to the pandemic; ER_CR = Emotional regulation (cognitive reappraisal); ER_ES = Emotional regulation (expressive suppression); SomAnx = Somatic anxiety; CogAnx = Cognitive anxiety; CovFear_P = Parents' Fear of COVID-19; SelfEf = Self efficacy; PanParent = Quality of parental pandemic practices.

Table reprinted from: Radanović, A., Micić, I., Pavlović, S., & Krstić, K. (2021). Pandemic Parenting: Predictors of Quality of Parental Pandemic Practices during COVID-19 Lockdown in Serbia. Psihologija, 54(3), 323–345. https://doi.org/10.2298/psi200731040r. Reprinted with the permission of authors.

N	Statement:	Answer
N1.	Parental distress due to the pandemic (Distress PD) causes Children's fear of COVID-19 (CovFear_C).	
N2.	Self efficacy (variable SelfEf) causes Parental distress due to the pandemic (variable Distress PD).	
N3.	Parental distress due to the national state of emergency (Distress NSE) causes Children' se fear of COVID-19 (CovFear_C).	
N4.	Distress PD and Distress NSE share more than 20% of variance.	
N5.	The correlation coefficients used here are contingency coefficients.	
N6.	We cannot conclude from the data that Distress CF and CovFear_C are correlated in the population.	
N7.	According to Taylor's recommendations, the correlation between Distress PD and CovFear_C can be considered to be moderate.	
N8.	There is no correlation between Distress PD and distress CF in the population.	
N9.	There is no correlation between SelfEf and Distress CF in the population.	
N10.	Quality of parental pandemic practices (PanParent) increases with the age of children (Age_Children).	

Let us now consider the answers:

L1 – false. We see that the correlation between these two variables is .034 and not statistically significant (correlations that are statistically significant are marked in the table with *, as we can read below the table), meaning that we accept the null hypothesis stating that the correlation in the population is 0 and treat that correlation as such.

L2 – true. We can see that although the correlation of -.052 is extremely weak, it is statistically significant (the sample of the study is very large!), so we reject the null hypothesis concluding that there is a non-zero correlation in the population and consider the correlation to be what the value of the correlation coefficient is. The statement indicates opposite directions of change of the two variables (one increases, the other decreases), which is a negative correlation and we can see that the correlation coefficient indeed is negative.

L3 – false. We can see that this correlation has the same value as the one discussed in L2, but the sentence here indicates the same direction of change of the two variables (one decreases, the other decreases also). The statement therefore states that the correlation is positive, while we can see from the value of the correlation coefficient in the table (-.052) that the correlation between these variables is actually negative, making this statement false.

L4 – true. Yes, the correlation between these two variables is positive and statistically significant (marked with *). It is .062. The statement describes the change of the two variables in the same direction, hence a positive correlation. So, the statement is true.

L5 – false. There is no indication in the table on which exactly correlation coefficient is calculated, but we can see that Safety has a negative correlation with Physical health and this would be impossible if Safety was a nominal variable. Correlation coefficients for nominal variables do not have a direction (in the way those for higher levels of measurement have). A side note – when reading a statistical text and we see that authors have calculated correlations, but did not report which coefficients exactly, most of the time we would be correct in assuming that those are Pearson product-moment correlation coefficients, if there are no other elements in the text to indicate that those would be inappropriate. This assumption is, however, not guaranteed to be 100% correct.

L6 – false. There is one correlation that is not significant – the one between Trust and Living conditions.

L7 – false. Cohen recommended that correlations around .1 be considered low. If we use the interpretation of his recommendation from this book stating that correlations below .2 should be considered low, we can see that this is false, as there are three correlation coefficients above .2.

L8 – true. Indeed, the correlation between Trust and Living Conditions is not statistically significant. Therefore, we accept the null hypothesis and consider these two variables not associated.

L9 – false. We can see that all presented correlations with Social relations are higher than with Living conditions.

L10 – true. The statement implies that the correlation is negative (lower values on one variable correspond to high values on the other variable). We can indeed see in the table that the correlation coefficient is -.086, statistically significant and negative. Hence the statement is true.

M1 – true. Taylor recommends that correlations between .36 and .67 can be considered moderate and the correlation between R and I is .45, therefore within the interval of moderate correlations.

M2 – false. Along the diagonal would be correlations of the variable with itself. Since every variable is identical with itself it would be the correlation between two sets containing the same values and that would be 1, not zero. It is, however, not displayed in the table, because calculating a correlation of a set of data and its identical copy is scientifically worthless.

M3 – true. The text below the table states that all correlation coefficients above .07 are statistically significant and we can see that there are no correlations in the table lower than this. Let us remember, that it is possible to simply set the threshold size of a correlation coefficient like this as long as sample size is the same for all considered correlation coefficients, such as would be the case if they were all calculated on the same sample. This is because the standard error of the correlation coefficient, when we are testing the null hypothesis that the correlation in the population is zero, is just a function of the sample size (1 divided by the square root of the number of entities in the sample). Since this is exactly what is done in the table – determining for which coefficients we can reject the null hypothesis, the threshold size of the correlation coefficient is the same for all coefficients in the table.

M4 – true. Cohen recommends that coefficients around .1 be considered weak. We interpreted that as correlations below .2. This one is .14, therefore weak, making the statement true.

M5 – false. Correlation between A and S, as we can read from the table is .48 and this is a statistically significant correlation and a pretty substantial one.

M6 – unknown. These are correlations. Correlation is not causation and we can make no inferences about causation from them. E might be the cause of C, but it might not be. We cannot tell just from what is available in the table.

M7 – true. Both variables increasing indicates a positive correlation and we can see from the table that their correlation of .64 is indeed positive.

M8 – true. Both variables decreasing together indicates the same direction of joint change, hence a positive correlation. We can see from the table that their correlation is .30, therefore positive.

M9 – true. Yes, the columns and rows of this table contain the same variables. And since these are Pearson's correlation coefficients, which are symmetrical, the correlation between variable A and variable B equals the correlation between variable B and variable A. In this case, for example the correlation between R and I is the same as the correlation between I and R. We should note, that if these were, for example, eta coefficients, or some other asymmetrical measure of association, this would not necessarily be the case.

M10 – false. No, they are Pearson's product-moment correlation coefficients. Although Spearman's rank correlation coefficients are a variant of the Pearson's correlation coefficient, it is customary in literature, if Spearman's rank correlations were calculated to write so and this is not the case here.

Table N – general comments. This is a somewhat more complex presentation of statistical data than the ones used so far, so we will provide this short explanation before proceeding to answers to statements. We can see that this table integrates data on variable means and standard deviations with correlations between variables. The first two columns with numbers are data on means and standard deviations of the variables whose names are given on the left. Each row is one variable, with some variables referring to certain properties of children (first two variables), while the other refer to properties of their parents. We can note that each of these variables is marked with a number. That number is used again in the column names in the upper part of the table, to denote those same variables, for the purposes of presenting a table of correlations between these variables. For example, we can see that the correlation between the variable SelfEf and the variable Age_Children is -.14. We can see that because the column in which we can see this value is named 2, referring to the variable also marked by the number 2 in the rows where it is named. In the same way, the column named 4 refers to the variable Distress CF, column 5 refers to Distress PD etc. Finally, asterisk (*) signs next to a correlation coefficient indicates that its significance value is lower than .05, i.e. that it is more statistically significant than .05. Two asterisks (**) indicate that the statistical significance value of the coefficient is also lower than .01 (i.e. that it is more statistically significant than .01). We can read this from the note below the table. Finally, readers should observe, that there is an extensive note below the table listing full variable names, while the table contains only abbreviations.

N1 – unknown. We can see that the correlation between these is positive and statistically significant (.37), so causation certainly is possible. However, correlation is not causation and we cannot know whether one of the variables causes the other or not.

N2 – unknown. We can again see that there is a negative correlation between these variables (-.15), a statistically significant one, making causation possible. However, correlation is not causation and whether one is the cause of the other remains unknown.

N3 – false. We can see that these two variables are not correlated i.e. that their correlation is almost zero and not statistically significant. This excludes that one causes the other as a causal effect would have to imply the existence of an effect and this would imply a correlation. Of course, it is still possible that both of these variables are part of some complex causal system, or that they could be causes of one another if suppressive effects of some other variable(s) were removed, but in the situation that these data refer to, this is not the case.

N4 – false. We remember that the proportion of shared variance is obtained by calculating the coefficient of determination i.e. the square of the correlation coefficient. Since we can see from the table that the correlation between these two variables is .21, the square of .21 (i.e. .21 × .21) is around .04, which corresponds to 4% and this is far lower than 20% mentioned in the statement.

N5 – false. Contingency coefficients cannot have negative values and these have negative values. Additionally, means and standard deviations were calculated meaning that these are not nominal variables.

N6 – true. The correlation, which is -.03 between these two variables is not statistically significant (almost 0 and no *).

N7 – true. According to Taylor, correlations between .36 and .67 are moderate and this correlation is .37, therefore moderate.

N8 – true. We can see that their correlation of .07 is not statistically significant. Therefore, we accept the null hypothesis that states that there is no correlation in the population.

N9 – false. We can read from the table that the correlation is .11. It is indeed low, but is marked with a *, meaning that it is statistically significant. Therefore, we reject the null hypothesis and conclude that there is a non-zero correlation in the population between these two variables.

N10 – false. We can read from the table that the correlation between these two variables is negative (-.27). Therefore, higher quality of parental pandemic practices corresponds to lower age of children, not higher.

References

Anđelković, V., Vidanović, S., & Hedrih, V. (2012). Relationship between Perceptions of Children's Needs Importance, Quality of Life and Family and Work Roles. *Ljetopis Socijalnog Rada, 19*(2).

Cohen, J. (1988). *Statistical Power Analysis for the Behavioral Sciences, 2nd Edition.*

Hackinger, S., Kraaijenbrink, T., Xue, Y., Mezzavilla, M., Asan, Van Driem, G., Jobling, M. A., De Knijff, P., Tyler-Smith, C., & Ayub, Q. (2016). Wide Distribution And Altitude Correlation of an Archaic High-altitude-adaptive EPAS1 Haplotype in the Himalayas. *Human Genetics, 135*, 393–402. 10.1007/s00439-016-1641-2

Hedrih, V. (2009). Profesionalna interesovanja i osobine ličnosti [Vocational Interests and Personality Traits]. *Godišnjak Za Psihologiju, 6*(8), 155–172. http://www.psihologijanis.rs/clanci/67.pdf

Hedrih, V. (2011). Provera konvergentne i diskriminativne validnosti analizom multiosobinske-multimetodske matrice na primeru PGI testa profesionalnih interesovanja zadatom uzorku iz Republike Makedonije [Assessment of convergent and discriminant validity through MTMM analysis on the example of the PGI vocational interest inventory administered to a sample from North Macedonia]. *Primenjena Psihologija, 4*, 393–408. http://primenjena.psihologija.ff.uns.ac.rs/index.php/pp/article/view/1138/1152

Hemphill, J. F. (2003). Interpreting the Magnitudes of Correlation Coefficients. *American Psychologist, 58*(1), 78–80.

Kruger, J., & Dunning, D. (1999). Unskilled and Unaware of It: How Difficulties in Recognizing One's Own Incompetence Lead to Inflated Self-Assessments. *Journal of Personality and Social Psychology, 77*(6), 121–1134.

Portugal, R. D., & Svaiter, B. F. (2011). Weber-Fechner Law and the Optimality of the Logarithmic Scale. *Mind&Machines, 21*(1), 73–81. 10.1007/s11023-010-9221-z

Radanović, A., Mićić, I., Pavlović, S., & Krstić, K. (2021). Pandemic Parenting: Predictors of Quality of Parental Pandemic Practices during COVID-19 Lockdown in Serbia. *Psihologija, 54*(3), 323–345. 10.2298/psi200731040r

Taylor, R. (1990). Interpretation of the Correlation Coefficient: A Basic Review: *Journal of Diagnostic Medical Sonography, 6*(1), 35–39. 10.1177/875647939000600106

7 Statistical tests for comparing two samples

In the chapter about inferential statistics, we discussed the procedures and concepts for making inferences about population values (parameters) based on sample values. These general procedures provide an overarching frame for making these inferences. However, when our goal is to compare specific properties of two different populations, to infer whether two samples come from populations that are equal in a specific statistical property or whether certain specific statistical parameters of two populations are in a certain hypothesized relationship, we need to use specific statistical tests, designed specifically for making the required type of comparison. While the concept of statistical significance or resampling confidence intervals and other appropriate procedures are used in all of these tests, the tests themselves include calculations that are necessary for these general statistics to be calculated.

So far, various statisticians have proposed and are still proposing various statistical tests, for testing various statistical hypotheses. The existing statistical tests can be used to compare two samples, to compare multiple samples at once, but also to compare a single empirical sample to a certain theoretical expectation expressed as either a specific distribution or a specific value of the statistic to be tested. The number of proposed statistical tests in literature is in the hundreds and likely in the thousands. However, the aim of this book being to introduce the reader to the basics of research statistics, we will focus in this chapter on the most commonly used and well-known tests for either comparing two samples or comparing a single empirical sample (i.e. sample consisting of real entities on which real measurements were performed, which were then included in the data matrix we are analyzing as data on the measured variables) to various theoretical expectations, either theoretical distributions or certain expected values of the parameter.

7.1 Paired samples and independent samples

An important issue to discuss at the start of a chapter on statistical tests is the difference between paired and independent samples. This is important because most statistical tests are intended for either paired samples or for independent samples and only a few have variants for both cases. And even with those that do have variants for both paired and independent samples these variants include somewhat different calculations that are just named the same test. There are statistical procedures that differ less than the variants of the same test for independent and for paired samples and are yet considered different tests! Whether the samples are paired or independent also dictates how the data matrix will be prepared i.e. how the data will be organized in a data matrix.

Paired samples are samples **where entities from different samples are paired in such a way that for every entity from one sample there is a corresponding entity**

DOI: 10.4324/9781003107712-7

from the other sample (or other samples in the case there are multiple paired samples!). This means that **for every single entity in one sample, we know precisely which entity from the other sample corresponds to it**. This correspondence of entities is established either by measurements being taken on entities that correspond to each other in some way or by multiple measurements of variable values on the same entities (and then the entity corresponds to itself). When hypotheses about the relations between these two samples are tested, calculations are then typically done in a way that makes comparisons between values of corresponding entities. Situations in research where paired samples are used include for example:

- **studies that involve two or more measurements of the same entities**, such as studies where measurements are taken before and after a certain procedure and longitudinal studies where measurements of the same entities are taken in a row. In these cases, **value of an entity from the first measurement corresponds to the value of the same entity in the second measurement** (and all the other measurements if there are more than one). If, for example, we tested the effects of a certain procedure on a group of people, the value of each person from the first measurement would correspond to the value of the same person in the second measurement (and all the subsequent measurements).
- **studies where comparable variables are measured on the same sample**. In this case again, the **value of an entity on one variable corresponds to the value of the same entity on the other variable**. An example of this are situations where attitudes towards certain objects are measured and then we pair measures of attitude of the same person towards different examined objects. Or situations where we compared achievements of a person on tests in different subjects (provided that they are first made meaningfully comparable in some way, for example by comparing positions on the distribution or in some other way!).
- **studies where different, but corresponding entities are used.** These include situations when the same variable was measured on romantic pairs or on family members or on a parent and a child or on teachers and their students etc. In such situations, values on a variable of one romantic partner would be paired with values of his/her romantic partner, or values of parents are paired with values of their children on the same variable or of people from the same family participating in the study or of teachers and their students.

In a data matrix, paired values are typically placed in the same row, the same as if they came from just a single entity, even in situations when they are actually from different paired entities.

Independent samples are **samples where entities from one sample do not have corresponding entities from the other sample** i.e. they are independent. Although the two samples might be, and usually are, comparable as wholes, there is no established correspondence between individual entities from one sample with individual entities from the other sample.

In a data matrix, two (or multiple) independent samples are typically represented as separate rows of data (separate entries) and a variable is typically included to indicate which entity (i.e. which row of data) belongs to which sample.

While there might always be exceptions as the field of statistics is constantly developing, **most typically statistical tests for paired samples require smaller differences between**

Table 7.1 An example of paired samples. A group of study participants (names given in the first left column) have rated the quality of services of 5 airline companies (here referred to as airlines A, B, C, D and E). Since these are all ratings on a 1-5 scale that are meant to be compared, we can treat the ratings of different airlines as comparable and make comparisons. For example, we could pair rating participants gave to Airline A to ratings these same participants gave to Airline B. In this comparison, Becky's rating of Airline A would be paired with her rating of airline B. Anita's rating of Airline A would be paired with her rating of airline B and the same would be done for all study participants. In the same way, we could pair ratings of all five airlines in the example with each other and compare them. The data is fictional.

How study participants rated airline companies, raw data Ratings were on a scale from 1 to 5, with 1 being the worst rating and 5 being the best

Study participant name	**Airline A**	Airline B	Airline C	Airline D	Airline E
Becky	**3**	5	4	3	2
Anita	**5**	5	4	5	5
Vladislava	**4**	2	5	5	2
Careen	**2**	2	1	2	2
Esmaeel	**5**	5	5	5	5
John	**1**	5	2	5	4
Dereck	**5**	4	4	5	4
Vladimir	**3**	1	3	5	4
Emmet	**1**	5	1	5	1
Mark	**5**	5	1	5	5
Ellen	**5**	5	3	5	5
Peter	**4**	4	2	4	4

Table 7.2 An example of independent samples. The study participants have been divided into two samples according to their gender. Please note the variable gender, where participants who declared themselves as females have been assigned number 1 and participants who declared themselves to be males have been assigned number 2. We have marked one group with bold font, while the other is written in normal letters. We can see, that we have a variable (Gender in this case), that contains information about the group each participant belongs to, but there is no correspondence between participants, no way to establish which participant from one group corresponds to which from the other, so there is no such correspondence. Independent samples can be of different sizes – we can see here that there are 5 females vs. 7 males in this example. The data is fictional.

Name	Gender	Neuroticism N	Extraversion E	Agreeableness A	Openness to experience O	Conscientiousness C
Becky	**1**	**45**	**55**	**55**	**65**	**70**
Anita	**1**	**22**	**67**	**50**	**72**	**62**
Vladislava	**1**	**37**	**25**	**45**	**65**	**45**
Careen	**1**	**55**	**32**	**65**	**42**	**35**
Esmaeel	2	25	25	32	28	60
John	2	52	12	42	35	28
Hamza	2	50	70	48	47	32
Vladimir	2	38	65	51	59	65
Emmet	2	40	25	65	61	65
Mark	2	25	30	22	45	50
Ellen	**1**	**32**	**45**	**45**	**68**	**65**
Peter	2	35	55	65	42	58

samples i.e. **smaller effect sizes to reject the null hypothesis or smaller sample sizes for the same effect size to reach statistical significance than tests for independent samples**. In other words, they typically have higher power compared to tests created for independent samples.

7.2 Comparing two means – t test

The t test is a parametric test created for determining **whether two samples come from populations with the same value of the arithmetic mean.** This is done by **testing the null hypothesis that the difference between means of two populations is zero**. This is based on the calculation of a standard error of the differences of the mean and then dividing the obtained difference between sample means with the standard error of difference between sample means. As with all standard errors, standard error of difference between means then depends on the variability of the samples on which they were calculated (i.e. their standard deviations/variances) and on sample size. The larger the sample and the smaller the variability of values of entities in both samples, the smaller the standard error of difference between means will be. The difference between means is than calculated and divided by this standard error of differences between means and the result obtained in this way is called the t statistic. The t statistic is than used to estimate the probability of obtaining the difference between means as was obtained between these two samples or larger when the difference between population means is 0 and this probability is the statistical significance of the t test. **If the result of the t test is statistically significant i.e. if p is lower than the accepted threshold of statistical significance (most commonly .05), we can reject the null hypothesis and conclude that samples do not come from populations with the same mean on the examined variable. On the other hand, if the value of statistical significance is above the accepted threshold, we accept the null hypothesis and conclude that there is insufficient evidence that population means are not the same (i.e. we will treat them as if they were the same).**

The **t test is a parametric** test meaning, in this case, that it assumes that the sampling distribution of the difference between means is normal. If we calculate the statistical significance of the t test by relying on the central limit theorem, we do not have a way of testing this assumption. However, the formulas for these calculations rely on the sampling distribution being normal. Since normality of the sampling distribution of differences between means cannot be tested, as we have just one sample and not the whole sampling distribution, **what researchers usually do is to test the normality of the data on the two samples and assume that if their distributions are normal that the sampling distribution of their differences is likely to be normal as well** (reminder: a sampling distribution is a distribution that would be formed by a large number of statistics obtained from a large number of samples drawn from the same population, or from the same two populations in the case of the sampling distribution of differences between means discussed here). However, Field (2009), for example, correctly notes that it is possible to have data from the two samples that is not normally distributed, but still have a normal sampling distribution of their differences. In any case, **a common research practice is that t test is used on the condition that the shape of the distribution of data from the two samples does not deviate too much from the theoretical normal distribution.**

Other important assumptions of the t test are also the assumption that the variances/ standard deviations of the two samples are equal and that all data come from separate/ independent measurements (this does not mean that samples should be independent – paired samples can also consist of independent measurements!). As for the first assumption about the equality of group variances, it should be noted that it is a sort of a default assumption for the t test, but there are also corrections to the formula for calculating the result of the t test for situations when variances of the two samples are not the same. One should only be careful to compare the variances beforehand and use the correction for the calculation in case they are not equal. The second condition, the one that measurements in the sample should be independent, means that inside each sample, each piece of data comes from a separate entity i.e. that there are no entities that were entered multiple times in the sample. If our entities are persons. for example, that means that each person should be represented in the sample only once. This should not be confused with situations where we have paired samples consisting of different sets of measurements coming from the same group of entities. This only means that within a single sample/group, on a single variable, there should not be multiple measurements of the same entities nor multiple copies of the same measurement.

Another way to test the null hypothesis of the t test is through resampling, typically meaning bootstrapping. In this procedure, **a sampling distribution is created by creating a predefined, large number of samples through sampling with replacement from the original sample and then defining a confidence interval that includes the desired percentage of the sampling distribution and is centered around the sampling distribution mean** (meaning that limits of the confidence interval are equally distant from the mean of the sampling distribution). After that, **the researcher checks whether 0 (or whatever other state is defined by the null hypothesis) is within the confidence interval or not. If it is within the confidence interval, we would than accept the null hypothesis and if it is not within the interval, we reject the null hypothesis.** It should be noted that this approach to testing the null hypothesis, essentially breaches the requirement of the t test that measurements be independent, as the procedure of sampling with replacement will necessarily create multiple copies of the same entity during the resampling process. However, this procedure is coming into increased use, at the moment this book is written, in spite of this.

The t test has variants for both paired and independent samples, meaning that it can be used for both types of samples. However, formulas for calculating the test statistics differ in these two cases. **The t test can also be calculated for only one sample, when we wish to test the hypothesis that the sample comes from a population with a specific value of the mean.**

While the results of the t test tell us whether to accept or reject the null hypothesis, rejecting the null hypothesis and concluding that means of the two considered populations are not the same, does not necessarily mean that the difference is sizable enough to be worth taking into account practically. Due to this, **it is customary to report a measure of effect size along with the results of the t test. Measures of effect size most commonly used along with the t test is the point-biserial correlation coefficient (r) and Cohen's d.**

The point-biserial correlation coefficient ($r_{p.bis}$) was already discussed in the chapter on correlations. It is calculated as a Pearson correlation between a (true) binary variable and an interval variable. In the case of the t test i.e. comparison of two samples, the binary variable is the variable containing information on which entity comes from which sample and the interval variable is the variable on which means are compared. The point-biserial correlation coefficient can also be calculated directly from the t statistic using a formula. As mentioned in the chapter on correlations, a useful recommendation for interpreting the size of correlations is the one proposed by Cohen (1988) stating that correlation coefficients of .1 can be interpreted as a weak/small, a coefficient of .3 can be considered moderate and a correlation of .5 can be considered large.

Cohen's d is a standardized difference between means. It is calculated by dividing the difference between means of the two samples with their pooled standard deviation. The pooled standard deviation is simply a mean of the two standard deviations (of the two samples) weighed by the sizes of the two groups. If the two groups consist of equal number of entities, then the pooled standard deviation is simply the mean of the two standard deviations. If the standard deviations of the two groups are the same, then it is that standard deviation. If the standard deviations are not the same and the two groups are not of the same size, then each standard deviation is multiplied by the number of entities in the sample from which it is calculated, these products (standard deviation x number of entities in its sample) are summed and the sum is then divided by the number of entities in both samples taken together. In this way, **Cohen's d tells us the size of the difference between the two means in units of standard deviations, in a way similar to z scores. Cohen suggested that d values of .2 can be considered small, a d value of .5 could be considered medium and d values of .8 and above could be considered large. Cohen's d values of less than .2 generally represent very small differences that can be considered negligible in most cases** and without very good theoretical justification for doing the opposite, even if the t test is statistically significant in such cases.

7.3 Comparing central tendencies of ordinal data on independent samples – the Mann-Whitney's U test and the Wilcoxon's sum-rank test

A popular non parametric alternative to the t test for independent samples is the Mann Whitney's U test. The U test compares the central tendencies of two independent samples. **What it effectively does is compare the mean ranks of two samples on the joint rankings. The null hypothesis is that the two samples come from the same population (or populations with equal properties) and that, due to that, when the data from both samples are ranked together, mean ranks of the two groups will be equal.** Some researchers like to say that in this way the U test is comparing medians of the two groups. However, the fact of the matter is that mean ranks are being compared and this need not always equal medians. It can be taken to be a test comparing medians if the shapes of distributions of the two groups are identical and this need not be the case. It can also be found in literature that it tests the hypothesis that there is a 50% chance that a randomly drawn value from one sample will be higher than a randomly drawn value from the other sample.

The U test is performed by first creating a joint ranking of both samples. This is done by joining the two samples together and then assigning rank one to the lowest value, rank 2 to the second lowest etc. When two entities have the same value, then they are assigned a mean of the two subsequent ranks. For example, if two entities have a value of 5 on the examined variable and this is the 10th lowest value in the sample, meaning that these two entities would occupy 10th and 11th place in the rankings, these entities would than both be assigned ranks 10.5 (because the mean of 10 and 11 is 10.5). **Such situations, when two entities have the same value are called ties and their ranks are then called tied ranks.** After this is done, the formula for calculating U is applied that is based on the number of participants in both groups and the sum of ranks in one of the groups. It is calculated for both groups and the smaller result is taken as the U test statistic. This statistic is than used in calculating statistical significance based on which we decide whether to accept or reject the null hypothesis.

The Wilcoxon's sum-rank test is similar to the U test and these two actually produce equal end results. The test statistic of this test is marked with W and it is the sum of ranks of the group with the smaller sum of ranks. The mean of the sum of ranks of the two groups is then subtracted from this statistic and the results is divided by the standard error of the W test statistic. Both of these later values are calculated from the number of entities in the two groups. **The result obtained in this way is a z scor**e (raw value (W) minus the mean (mean sum of ranks) and divided by the standard deviation (standard error is the standard deviation of the sampling distribution)) **and indicates the position of our score on the sampling distribution we would obtain if the populations the two groups come from were equal. We can than compare that z value to known threshold z score values to determine if the W statistic is statistically significant (that is 1.96 if we want a two tailed significance with a .05 threshold).**

While these calculations might sound complex, the basic idea behind both tests is that, if the two groups come from equal populations, when you create a joint ranking of both groups, there will roughly be equal numbers of cases with high and with low ranks in both groups. If high scores tend to be concentrated in one group and low in the other, than the two groups are not equal. For example, imagine two teams running a race and us recording the order in which they passed the goalpost i.e. their ranks. If the two teams were equally fast, then there will be people who finished among the first and people who finished among the last in both groups. If we notice that people from one team tended to generally reach the finish line before people from the other group i.e. that they mostly occupied the first part of the rankings, while the second team mostly occupied the second part of the rankings, we would conclude that one team is faster than the other.

When interpreting the results of both the Mann-Whitney's U and the Wilcoxon's sum rank test, the most important statistic for the purpose of interpretation is, as with the other tests, statistical significance. Again, if it is above the accepted threshold (usually .05), we accept the null hypothesis, concluding that there is no evidence that the two samples come from different populations and we thus treat them as coming from the same population. If it is below the threshold (usually below .05), we reject the null hypothesis and conclude that the two samples do not come from the same population or from populations with the same properties.

While it might still be relatively unclear which exactly statistics do these two tests compare, their results undoubtedly refer to the central tendencies of the populations from which the two samples are drawn. What we can see in literature is that researchers often tend to use the results of the U test as if they referred to the relationships between means (although they usually avoid writing that directly).

Both the Mann-Whitney U and the Wilcoxon's sum rank test require that the data be at least on the ordinal level of measurement, but, as they are **nonparametric**, do not have any requirements about the shape of the distribution of data.

The effect sizes that can be used with these tests are different. In the literature, we can often find the **rank-biserial correlation** coefficient used as an effect size measure along with the Mann-Whitney U test. **The rank-biserial correlation coefficient is an indicator of association between one binary and one ordinal variable.** One can also find the **Spearman's rank correlation coefficient** (which is very similar to the rank-biserial correlation coefficient) calculated between the binary variable indicating which entity is from which sample and the variable on which the samples are compared **used as a measure of effect size.** Also, Field (2009) in his book cites Rosenthal and suggests that the z score obtained in the Wilcoxon's test procedure be converted to a correlation coefficient by dividing it with a square root of the total sample size. Given that these are all correlation coefficients, their interpretation is than the same as with all the other correlation coefficients.

7.4 Comparing central tendencies of paired samples – the sign test and the Wilcoxon's signed rank test

The **most well-known alternatives to the t test for paired samples are the sign test and the Wilcoxon's test.**

The **sign test tests the null hypothesis that two paired samples come from populations with equal medians.** It is performed by **setting the data from the two samples in two parallel columns,** so that corresponding pairs of data from the two samples are in the same row. Then, **for each entity from one sample we record whether the corresponding entity from the other sample has a value that is higher, lower or equal to it.** If it is higher, it is typically denoted by +, if it is lower, it is denoted by −, and if the values are equal that relationship is denoted by =. The expectation is that, **if the null hypothesis is true, probabilities for + and − will be equal i.e. the total number of + and the total number of - signs will be roughly equal. The larger the difference between the number of + and − signs, the more statistically significant the results of the sign test will be.** For example, imagine that we had pairs of students formed according to their academic ability and motivation to take a course in statistics. Then we divide them into two groups by putting one student from each pair of students into one group and the other one from the pair into the other. Then we have one group of students take one course in statistics and the other group the other. We maintain the data about which student from one group is paired with which student from the other group making these two samples paired. Since the students are paired, the "quality" of both groups of students are the same relative to their ability to make use of the statistics course. Given this, if both courses in statistics were of the same quality, we would expect that after the course the statistics knowledge of both groups will be equal. That means that when we compared the values of students from the two samples, by comparing each one with his/her pair, we should expect that there be an equal number of situations where a student from the first group performed better and the number of situations where the student from the second group performed better. If these numbers are roughly the same i.e. if there are clearly more cases of students from one group being better than cases from the other group being better, this could be taken to mean that the quality of the two courses is not equal, provided we have reasonably excluded any other factor that could have caused the difference between groups. This is the so called experimental study design without a pretest and we could use the sign test to make inferences about its results if it turned out that we cannot use the t-test for paired samples for any reason.

There is a bit of an ambiguity about the required level of measurement for the sign test in the literature. The **minimum level of measurement required for the sign test is the ordinal level of measurement**. However, a fact of the matter is that if the data represent two separate rankings, with each of the two paired samples containing ranks from 1 to the number of entities in the sample, the sign test will not be useable, because the number of + and − signs will be equal. However, if the values are created by jointly ranking both samples or if any other measurement or assessment system is used that puts both samples in the same frame of reference, that would be sufficient for performing the sign test. **The sign test poses no requirements about the distribution of data.**

The **Wilcoxon's signed rank test** is **based on the same concept as the sign test, with the differences being that it takes the size of the difference between corresponding values into account**. It is also a nonparametric alternative for the t test, but however, unlike the sign test, **its results are fully meaningful when the data are on the interval level of measurement.** This is because the calculation of this test, requires calculation and comparison of differences between value and for this to be fully meaningful, a fixed unit of measurement is needed.

The setup for the Wilcoxon's signed rank test is the same as for the sign test, but the calculations are a bit more complex:

•   Data are organized into two parallel columns with corresponding values being in the same rows, just as with the sign test.
•   The differences between the corresponding values are calculated. These differences may be positive, negative or zero depending on which value in a pair is larger or if they are equal.
•   In the next step, positive and negative signs are removed and all differences are ranked according to size (therefore – regardless of whether the difference is positive or negative) – the smallest difference is assigned rank 1, the second smallest rank 2 and so on. If two or more pairs of values have the same difference, than they are both assigned the same rank that is the mean of the place they would be taking on the rankings. For example, if there are three identical values that would be taking ranks 4, 5 and 6, then all of the three pairs are assigned rank of 5 because $(4 + 5 + 6)/3 = 5$.
•   Then, the signs from step 2 are assigned to these ranks. In this way we obtain the **signed ranks**. For example, if a difference between two paired values received a rank 6 and the second value was larger than the first, it becomes a -6. If the first value was larger, it becomes 6 (it is actually +6, but we do not usually write the sign with positive values).
•   The signed ranks are summed and, based on this result, statistical significance of the test is calculated.

The null hypothesis of the Wilcoxon's signed rank test is that the two paired samples come from the same population or from populations with the same central tendency. When the null hypothesis is true, the sum of these signed ranks is expected to be around 0. The more it differs from 0, the more statistically significant the result of the Wilcoxon's signed rank test becomes (i.e. the statistical significance value becomes smaller).

As with all other tests, **when the value of statistical significance is lower than the accepted threshold (which, most commonly, is .05) we reject the null hypothesis** and conclude that the central tendencies of the populations the two samples are from are not the same. If the statistical significance value is higher than the accepted threshold, we accept the null hypothesis and conclude that there is no evidence that central tendencies of the populations the two samples are from are different.

Measures of effect size that could be used along with the sign test and the Wilcoxon's test include the rank-biserial correlation coefficient, but also the Spearman rank correlation coefficient can be found in use. Some authors also prefer the use of **Somers' D**, which is an effect size measure based on the comparison between the number of pairs where the first value is larger than the second and the number of pairs where the first value is smaller than the second i.e. comparison between the number of + and the number of − pairs (referred to as concordant and discordant pairs in the calculation). **Somers' D ranges from –1 to 1 based on whether there are more + or more − signs.** For example, when all pairs have − signs i.e. when the value from the second sample is always larger than the first, D will be -1. However, similar to the interpretation of the point-biserial correlation coefficients, this sign should not be interpreted as having any more substantial or permanent meaning except indicating

Table 7.3 An example of the calculation of the sign test and of the Wilcoxon's signed rank test. The first three columns represent the data, the fourth from the left represents the calculation of the sign test, while the four columns on the right represent the setup for the calculation of the Wilcoxon's signed rank test. If the null hypothesis is true the number of + and − signs of the sign test will be roughly equal. If the null hypothesis of the Wilcoxon's signed rank test is true, the total sum of ranks should be 0. The data is fictional

The data – values of pairs on the two paired samples			The sign test	The Wilcoxon's signed rank test			
Pairs of study participants (person in group 1 – person in group 2)	Test scores group 1	Test scores group 2	Sign of difference between groups 1 and 2 (1-2)	The difference between values in 1 and 2	The difference without the sign / absolute difference	Rank of the absolute difference	Signed rank
Becky – Anna	45	55	−	−10	10	2	−2
Anita – Serena	22	67	−	−45	45	11	−11
Vladislava – Juliane	37	25	+	12	12	3	3
Careen – Vanya	55	32	+	23	23	8	8
Esmaeel – Chen	25	25	=	0	0		
John – Yuri	52	12	+	40	40	10	10
Hamza – Damyean	50	70	−	−20	20	6.5	−6.5
Vladimir – Donald	38	65	−	−27	27	9	−9
Emmet – William	40	25	+	15	15	5	5
Mark – Victor	25	30	−	−5	5	1	−1
Ellen – Michaela	32	45	−	−13	13	4	−4
Peter - Philip	35	55	−	−20	20	6.5	−6.5

Sign test results: Total number of - : 7, total number of + : 4

Wilcoxon' signed ranks results: Sum of negative ranks: -40 ; Sum of positive ranks: 26. Total sum of ranks: -14

whether the first or the second group tend to have higher ranks. We should keep in mind that which group is the first and which one the second is arbitrary and that a different designation of groups would result in a different sign of the D.

7.5 Comparing standard deviations/variances – Levene's test

A test commonly found in literature when **comparison of variances is needed is the Levene's test. The null hypothesis of the Levene's test is that all compared groups come from populations with equal variances i.e. that the differences between population variances are 0.** Yes, unlike the other tests presented in this chapter and contrary to the title of this chapter, Levene's test can compare more than two groups simultaneously. As with all other tests, **when the results of the Levene's test are statistically significant (i.e. when the value of statistical significance is lower than the accepted threshold, which is most commonly .05), we reject the null hypothesis, concluding that the compared groups do not come from populations with equal variances.** When the test is used to compare two samples, this means that the variances of the two populations these two samples come from are different. However, when we are comparing more than two samples, a statistically significant result of the Levene's test means that variances of the populations in questions are not all the same, but since there are more than two, we do not know which population likely has a different variance compared to which other population. A statistically significant result of the Levene's test used to compare multiple groups does not by itself indicate that variance of every group's population differs from the variance of every other group's population. It only indicates that they are not the same and the more groups there are that are compared with the test, the larger the number of things this might mean precisely. It might of course mean that variances of all considered populations are different from each other, but it might also mean that only two of the compared populations have likely different variances or that one of the compared populations differs from all the others, which are equal. However, based solely on one calculation of the Leven's test on multiple groups, we do not know which of these is the case. To find precisely which sample's population variance likely differs from which other sample's population variance, we would need to rerun the Leven's test to make pairwise comparisons of the samples i.e. compare them two by two using this test. On the other hand, **a result of the Leven's test that is not statistically significant means that we should accept the null hypothesis and therefore conclude that there is no evidence that populations for which the samples are representative have different variances.**

Levene's test is a parametric test, meaning that it assumes the normal sampling distribution and that the **data are at least on the interval level of measurement.** Also, given that a standard deviation is the square root of the variance, **the results of Levene's test equally refer to the relationships between standard deviations of the populations of the compared samples.**

Levene's test can be used to test research hypotheses about the differences in variability of populations from which samples are, but **most commonly we will find it used prior to other statistical tests that rely on the assumption that the variances of populations the compared samples are from are equal i.e. that rely on the assumption of the homogeneity of variances.** Such is the situation with the t test described in this book. Levene's test is often performed prior to calculating the t test to test whether the variances of populations the two samples come from are equal. **If Levene's test turns out to be statistically significant, then it is concluded that variances of populations compared groups are from are not equal and t test is performed without the assumption of the homogeneity of variances, which implies using a correction in the way statistical significance of the t test is calculated.**

A problem with using Levene's test to assess the homogeneity of variances prior to using another test happens when the sample size is very large. As with all calculations of statistical significance, as the sample size gets larger, the standard error gets smaller. Due to this, when sample sizes become very large, very small differences between compared values start becoming statistically significant. In this case, this would mean that with large samples we will often encounter situations where Levene's test is statistically significant, but the differences between variances of the groups are practically negligible. In such situations, we can simply acknowledge the situation and conclude that differences between variances are too small to prevent us from accepting the assumption of homogeneity (of variances), but a better and more objective approach would be to use the **variance ratio**, which is the **ratio between the largest and the lowest compared variance**. The critical values of the variance ratio depend on the number of entities in the sample and the number of samples being compared and can be found for example in Field (2009).

7.6 Comparing two distributions – Kolmogorov-Smirnov test, chi square, Wald-Wolfowitz test

Probably the most popular test for comparing two distributions is the Kolmogorov-Smirnov test or K-S test for short. **The Kolmogorov-Smirnov test tests the null hypothesis that two samples come from populations with equal distributions**. This test can be used to compare two empirical distributions, but, in scientific literature, we will much more often find it used to compare an empirical distribution to a certain theoretical distribution (with the same mean and standard deviation as the considered empirical distribution) i.e. to compare the shape of the distribution obtained on the study sample with the shape of a certain theoretical distribution. Again, the most commonly used theoretical distribution for this type of comparison found in literature is the normal distribution. In other words, **we will most commonly find K-S test used as a test of normality of the sample distribution** i.e. **used to test the null hypothesis that the study sample comes from a population that has a normal distribution on the examined variable.** This is most often done in preliminary examinations of empirical data, when researchers decide whether they can use parametric tests i.e. tests that rely on the assumption that the data is normally distributed or that the sampling distribution is normal (and then the normality of the data distribution is used as a proxy for this, since the sampling distribution is unavailable for testing).

The Kolmogorov-Smirnov test requires that the data be at least on the interval level of measurement. It also requires that the distribution data is compared to must be (however, there is also a version that works with discrete distributions) **fully specified**. The test is performed by calculating a series of differences between the two distributions and the K-S test statistic, called **Kolmogorov-Smirnov D**, is based on the difference between the two distributions at the point where that difference is the largest. This means that for two distributions that differ in all their parts and for two distributions that are equal everywhere except at one part, the K-S test results will be the same as long as that largest difference between the compared distributions is the same.

When the K-S test is statistically significant i.e. when the value of statistical significance is below the accepted threshold (usually .05), we reject the null hypothesis and conclude that the compared samples do not come from populations with equal distributions (when two samples are compared) or that the sample does not come from a populations with the shape of the distribution equal to the theoretical shape of the theoretical distribution with which it was compared. In the most common case of K-S test use, when it is used to compare the sample distribution with the normal distribution i.e. **as a test of distribution normality, a statistically significant K-S test indicates that the sample does not come from a population with the normal distribution on the examined variable.** On the other hand, **when the results of K-S test are not statistically significant, i.e. when the value of statistical significance is above the acceptance threshold (usually .05) we can conclude that there is no evidence that the distribution of the population is not normal and can hence be treated as normal**. We should also be aware that, as with all calculations of statistical significance, as sample size increases, so do the differences that become statistically significant decrease. This means that with **very large samples, even very small deviations from the normal distribution will result in a statistically significant K-S test.** Due to this, **when a sample is large and we want to decide whether to use tests that rely on the normal distribution of data in further analysis it is much more reasonable to rely on statistics showing the level of deviation from the normal distribution – skewness and kurtosis than on K-S test**, because, with sufficiently large samples, completely negligible deviations from normality will still result in a significant K-S test.

The Pearson chi-squared test, often referred to in literature as just **chi square or the chi square test** is **a procedure for comparing an empirical distribution with a theoretical distribution that can be presented in a form of a contingency table** i.e. in the form of a set of discrete categories. This is done with the idea of testing the hypothesis that the sample comes from a population that has the distribution of the same shape as the theoretical distribution used for comparison. The test is performed by calculating the frequencies of each category of the empirically assessed variable i.e. counting how many entities there are in each category. Frequencies obtained in this way are called the observed frequencies. Based on the total number of entities in the sample, another distribution of frequencies across categories is created that is in line with the theoretical distribution we want to compare with our empirical distribution. In other words, we estimate what would the frequencies of each category be if the variable had the theoretical distribution we are examining (i.e. how many entities would be in each category). The frequencies created in this way are called expected frequencies and the whole distribution is called the expected distribution. The test statistic called the chi square statistic is then calculated based on the sum of squares of differences between these two sets of values. The larger the differences between the observed and expected distributions and the more categories there are that differ, the larger the chi square statistic will be and thus more significant (for the same sample size!). **The null hypothesis of the chi square test is that the sample comes from a population whose distribution is the same as the expected distribution. When chi square test is statistically significant, we reject the null hypothesis and conclude that the distribution of the population the sample is from does not equal the expected**

distribution. If it is not statistically significant, we conclude that there is not enough evidence that the population distribution differs from the theoretical distribution we based our expected frequencies on.

One important feature of the chi square test is that **it can be used on nominal data** (it works with categories!), **but chi square test calculations are also made for originally continuous data that has been converted to a certain fixed number of categories**.

The **most commonly used theoretical distribution to compare with an empirical one using chi square is the uniform distribution**, but chi square can be used to compare the empirical data distribution with any theoretical distribution as long as it can be used to create expected frequencies. It can also be used to test the normality of the distribution and, in that role, it differs from the K–S test in the fact that all differences between the compared distributions affect the result of the chi square test and not just the largest one, as is the case with K–S.

Another common use of the chi square test is to test the null hypothesis that two nominal (or at least categorical) variables come from populations in which they are not associated. This is done by first creating a crosstabulation of the two variables i.e. categories that are defined by both variables. For example, if the two variables were gender and hair color and gender had categories male, female and other and color had categories black, brown, blue and other we would create a category for every possible combination of the two values. We would therefore have a category for males with black hair, females with black hair, other with black hair, males with brown hair, females with brown hair and so on until all possible combinations are exhausted. We than count the frequencies of entities in each of these (combined) categories in the study sample and these are the observed frequencies. Finally, we make an estimation of what would the frequencies in a sample of our size be for each category if the two variables were not associated. These estimates serve as expected frequencies for calculating the chi square statistic. **In case of using the chi square to test for association between variables, rejecting the null hypothesis (i.e. when the chi square is statistically significant) indicates that the two variables are likely associated in the population. Accepting the null hypothesis then indicates that there is not enough evidence for the association of the two variables in the population, which means that we should be treating them as if they were not associated in the population.**

One requirement of the Pearson's chi square test is that the observations i.e. data be independent. This means that chi square is not an appropriate statistic when we have multiple measurements taken from the same entity. Another important **requirement of this test is that there are no expected frequencies lower than 5.** When this happens, the researchers should either consider merging categories, if that can be done meaningfully, to get all expected frequencies above 5 or use some alternative test that is appropriate for samples that small. For example, one often mentioned alternative to the Pearson's chi square test when the goal is to test the association between two nominal variables and the sample is small is the **Fischer's exact test**. Also, when Pearson's chi square is used on 2x2 tables (e.g. association between two binary variables), it is suggested (e.g. Field, 2009) that it tends to produce too small significance values and in such cases the application of the **Yates's continuity correction** is proposed. This correction effectively reduces the size of the difference between each pair of expected and observed frequencies before they are squared in the calculation of the chi square statistic, thus reducing the size of the chi square statistic.

Various different statistics can be used as **measures of effect size** along with the Pearson's chi square test. When the chi square test is used to test for the association of two nominal variables, correlation coefficients for nominal variables can be used as adequate effect size measures. These include the **contingency coefficient, phi coefficient and Cramer's V**. These coefficients can themselves be derived from the value of the chi square statistic or include similar data in the calculation. One more possible effect size measure that can be used with the chi square is the **odds ratio. Odds ratio is simply the ratio between two odds, showing how many times odds of one outcome are larger than odds of another outcome**. It is obtained by dividing the two odds. Odds ratio is the easiest to interpret when we are dealing with a 2x2 frequency table. For example, if we had one variable that registered whether a study participant is a high school or a primary school student and another showing whether the student passed or failed a certain test, we could calculate the odds of a high school student passing the test and then the odds of a primary school student passing that test. We would than divide these two odds to show us, for example, how many times more likely a high school student is to pass the test than a primary school student.

In a situation when we have only **a single binary variable and want to test whether the sample came from a population with a certain proportion of cases in each of the categories** (for example, a population in which both categories are equally probably or where there is a certain known probability for each of the two categories), **the test to be used for this is the binomial test**. It **tests the null hypothesis that the sample comes from a population with the predefined probabilities of the two categories**. This test is exclusively intended for making inferences about population proportions of two categories of a single binary variable.

The **Wald–Wolfowitz test** is a test **for comparing distributions of two independent samples** that can be applied to **data that are at least on the ordinal level of measurement. It tests the null hypothesis that two independent samples come from populations with identical distributions**. The assumption is that distributions are identical both in shape and central tendency i.e. that they are completely overlapping. The premise the test is based on is that if we have a joint ranking based on values of entities from both samples, if the two distributions are identical, the entities from the two groups will be mixed on the rankings i.e. **there will not be many uninterrupted series of entities from a single group** as we go down the list. In other words, as we go from the start of the list, there will be one or a couple of entities from one group, then one or a couple of entities from the other, than again from the first and so on. On the other hand, for example if all entities from one group have higher values than entities from the other group, which is the most extreme difference detectable by this test, when we look at the list from the beginning, we will first see a series of all entities from one group and then it will be followed by a series of entities from the other group. The Wald–Wolfowitz test is performed by starting from the beginning of the rankings list and then counting the number of uninterrupted series of entities from the same group. The number of such series is then compared to the maximum possible number of such series and the statistical significance of the test statistic obtained in this way is calculated. When the two groups are of equal size, the maximum number of series equals the number of entities (a list where after each entity from one group comes an entity from the other group). However, when the two groups are not of equal size, the maximum number of series is 2 × the number of entities in the smaller group + 1. This is because even when each entity from the smaller group is preceded and followed by an entity from the larger group, there will still be entities from the larger group being next to each other on the list simply because there are not enough entities from the smaller group to interrupt all series from the larger group, because there are less of them. One can imagine this like a task of braking up a single line with a certain predefined number of dots. For example, if we had 4 dots to place them along a single line, how many parts can we divide that line into, using these 4 dots? The answer is of course into 5 parts. If we imagine that the dots are actually individual entities from a smaller group we arrive to our formula for the maximum number of series when the groups are unequal – 2 × 4 + 1= 9.

Anyway, the larger the difference between the maximum possible number of uninterrupted series and the observed number of series, the more statistically significant the Wald–Wolfowitz test will be. **When the statistical significance of the Wald–Wolfowitz test is below the accepted threshold for statistical significance (usually .05) i.e. when the result is statistically significant, we reject the null hypothesis and conclude that the two compared samples do not come from populations with identical distributions.** However, **when the statistical significance of the test is above the accepted threshold for statistical significance i.e. when the test is not statistically significant, we accept the null hypothesis and conclude that there is not enough evidence that the samples come from two different populations and proceed to treat their distributions as equal.**

One important thing to note is that the Wald–Wolfowitz test **works best when there are no situations when entities from different groups have the same value i.e. when there are no tied ranks of entities from different groups. When there are many tied ranks, the test becomes unusable.** This is because in situations where there are multiple entities from different groups with the same values, we have to decide on how many uninterrupted series there are in such situations. For example, if we have 10 entities with the same value on the examined variable, this means that they would all be sharing the same rank. Let us imagine that of these 10 entities, 5 are from one group and 5 from the other. We can than consider that these 10 entities form just 2 series of values – first the 5 from one group and then 5 from the other group i.e. 2 series. Or we can, with equal justification, consider that they constitute 10 series – one from the first group, then one from the second, then one from the first group, then one from the second and so on up to 10. Given that all these entities share the same value, both of these interpretations are equally valid. However, in one case we will have just 2 series out of these 10 entities, thus substantially contributing towards making the test result statistically significant, and in the other case, there will be 10 series out of these 10 cases, bringing the result of the test as a whole away from reaching statistical significance. While such ties involve a small part of the sample, the difference between the test results if we go with the first option is not much different from the test outcome if we went with the second option. However, when the number of such ties increases to include a sufficiently large part of the sample the test becomes unusable because the conclusions that would be made based on one and the other option for deciding on the number of series become completely opposite in that way.

Table 7.4 Examples of the various numbers of series calculated in the scope of the Wald-Wolfowitz test and representation of how the number of series is calculated. Example 1 is the minimum number of series possible, a situation when the test results would achieve the highest possible statistical significance (i.e. the lowest possible p value), that is the highest possible difference from the state postulated by the null hypothesis. Examples 2 and 3 represent situations postulated by the null hypothesis – the maximum possible number of series indicating total overlap between distributions, with the difference that example 2 is the situation when groups are of equal sizes and group 3 is an example of this situation when groups are not of equal sizes. Example 4 represents a situation when there is some overlap between distributions, but it is not total i.e. the number of series is somewhere between the minimum and the maximum possible. None of the examples represent a situation with tied ranks, when the test would produce two counts of the number of series, one if counting the entities with tied ranks as mixed and the other counting them as not mixed

Ranks from the joint rankings of the two groups to be compared (setup for the Wald-Wolfowitz test)	There are two groups – 1 and 2 and the numbers in these columns represent the group the entity that achieved the rank indicated by the row is from (left in each example) and the count of the number of series of entities from the same group (right in each example).						
	Example1 – no mixing between groups, one group clearly higher, distributions completely distinct		Example 2 – perfect mixing between groups, distributions completely overlapping		Example 3 – perfect mixing between groups but group 1 is 3x smaller than group 2, distributions completely overlapping		Example 4 – some mixing between groups, but not perfect, unequal groups, distributions overlapping somewhat
	Group membership	Series count	Group membership	Series count	Group membership	Series count	Group membership	Series count
1	1	1	1	1	2	1	2	1
2	1	1	2	2	2	1	2	1
3	1	1	1	3	2	1	2	1
4	1	1	2	4	1	2	2	1
5	1	1	1	5	2	3	1	2
6	1	1	2	6	2	3	1	2
7	1	1	1	7	1	4	2	3
8	1	1	2	8	2	5	1	4
9	1	1	1	9	2	5	1	4
10	1	1	2	10	1	6	1	4
11	2	2	1	11	2	7	2	5
12	2	2	2	12	2	7	2	5
13	2	2	1	13	1	8	1	6
14	2	2	2	14	2	9	1	6
15	2	2	1	15	2	9	2	7
16	2	2	2	16	2	9	1	8
17	2	2	1	17	1	10	2	9
18	2	2	2	18	2	11	2	9
19	2	2	1	19	2	11	2	9
20	2	2	2	20	2	11	2	9
The total number of series	2		20		11		9	
The maximum possible number of series given the sizes of groups	20		20		11		17	

7.7 Let us apply what we learned so far!

Let us now try to apply what we have covered in this chapter through a couple of exercises. Please refer to the start of the book for the general instruction for completing the exercises. Our suggestion is that you first read the excerpt and the statements and provide your own answer. You may write it in the Answer column, and after that look up the answers and compare your own answers with them.

Exercise O. Statistical tests. (Rakić-Bajić & Hedrih, 2012)

Gender	Mean	SD	t	p	Effect size (r_{pbis})
males	40.03	16.03	2.196	.03	.171
females	35.08	12.27			
Marital status	Mean	SD	t	p	Effect size (r_{pbis})
married	33.14	13.01	−2.193	.03	.159
not married	38.15	13.99			
	Dependent variable – Excessive use of the Internet				

Table is based on data from a study published in Rakić-Bajić & Hedrih (2012)

O	Statement:	Answer
O1.	There is a statistically significant difference in means of males and females in the Excessive use of the Internet.	
O2.	There is a statistically significant difference in means of persons who are married and those who are not in the Excessive use of the Internet.	
O3.	The size of the difference between males and females on the Excessive use of the Internet is high.	
O4.	The size of the difference between mean values of persons who are married and those who are not on the Excessive use of the internet is medium.	
O5.	The test used here is parametric.	
O6.	Excessive use of the Internet is on the nominal level of measurement.	
O7.	There is a substantial difference between standard deviations of males and females on the Excessive use of the internet in the population.	
O8.	The difference between standard deviations of persons who are married and those who are not married on the Excessive use of the internet is not statistically significant.	
O9.	It was possible to use the U test here for a similar purpose instead of the t test.	
O10.	Excessive use of the internet is more pronounced in males than in females.	

Exercise P. Statistical tests. (Čižman Štaba et al., 2021)

Differences scores between groups (passed the driving test vs. did not pass the driving test or passed it with adjustments) on different tests and demographic variables

		N	U	df	p	r
Demographic variables	Age	63	461	61	.780	0.04
	Education (in years)	63	587	61	.119	0.20
	Amnesia (in weeks)	33	74	31	**.027**	0.39
	Time from injury (in months)	63	285	61	**.006**	0.35
	Coma duration (in weeks)	40	82	38	**.001**	0.50
	GCS	40	317	38	**.001**	0.51
Mediatester variables	RTAV (*t*-value)	63	729	61	**< .001**	0.44
	ART Total RT (in seconds)	62	386	60	**.272**	0.14
	CRT – correct reactions – (M in seconds)	63	361	61	.092	0.21
	18 LRT (in seconds)	63	315	61	**.020**	0.29
	VRT Total RT (in seconds)	62	404	60	.357	0.12
Neuropsychological assessment variables	TOL – Total Move Score	63	385	61	.180	0.17
	CVLT (sum of all recalled words in all the trials)	63	629	61	**.039**	0.26
	HOVT (sum)	63	599	61	.100	0.43
	D2	63	228	61	**< .001**	0.46
	CTMT (composite index – T)	63	743	61	**.000**	0.45
	COG (*t*-value)	63	637	61	**.029**	0.27
	LVT (*t*-value)	63	665	61	**.009**	0.33
Driving-related variables	Average of all the ratings from the instructor	47	458	45	**< .001**	0.58

Note. The results where *p* < .05 are shown in bold. The variables included in the final regression model are shown in italic (GCS, 18 LRT, CTMT).
N = number of participants; *U* = Wilcoxon-Mann-Whitney U statistic; *df* = degrees of freedom; *r* = size of effect measure.

Table reprinted from: Čižman Štaba, U., Klun, T. R., & Robida, R. (2021). Predicting Factors of Driving Abilities after Acquired Brain Injury through Combined Neuropsychological and Mediatester Driving Assessment. Psihologija, 54(2), 137–154. https://doi.org/10.2298/PSI200408024C. Reprinted with the permission of authors.

P	Statement:	Answer
P1.	Those who manage to pass the driving test after a traumatic brain injury tend to be older than those who do not.	
P2.	The difference between the two groups on RTAV is statistically significant.	
P3.	The difference between the compared groups is higher on 18 LRT than on RTAV.	
P4.	The effect size measure is the rank-biserial correlation coefficient.	
P5.	Driving skills directly cause the differences between groups in values of D2.	

(Continued)

P	Statement:	Answer
P6.	The median significance score is high only for the variable D2.	
P7.	The statistical test used here to compare the two groups is a parametric test.	
P8.	The two groups come from populations of roughly equal average age.	
P9.	We should reject the null hypothesis of this test for the variable Coma duration (in weeks).	
P10.	We should accept the null hypothesis of this test for the variable GCS.	

Exercise Q. Statistical tests

			Marital status		Total
			married	not married	
Education	University	% within education	42.7%	57.3%	100%
		% within marital status	83.7%	39.6%	51.1%
		% of total	21.8%	29.3%	51.1%
	Secondary school	% within education	8.2%	91.8%	100%
		% within marital status	14.3%	56.1%	45.2%
		% of total	3.7%	41.5%	45.2%
	Other	% within education	14.3%	85.7%	100%
		% within marital status	2.0%	4.3%	3.7%
		% of total	.5%	3.2%	3.7%
	Total	% within education	26.1%	73.9%	100%
		% within marital status	100%	100%	100%
		% of total	26.1%	73.9%	100
		Statistic name	Statistic value	Statistical significance	
		Pearson Chi Square	28.325	p<.001	
		Cramer's V coefficient	.388	p<.001	
		N	188		

The data is from the authors own research study.

Q	Statement:	Answer
Q1.	The association between marital status and education level is not statistically significant.	
Q2.	The intensity of the association between marital status and education level is moderate according to Cohen's recommendations.	
Q3.	The variable "education level" is on the nominal level of measurement.	
Q4.	Secondary school education is more common among the married participants in the sample than among the unmarried ones.	

(*Continued*)

Q	Statement:	Answer
Q5.	There are more than 200 participants in the sample.	
Q6.	The correlation coefficient that was calculated here can have negative values.	
Q7.	The mode of the variable Education on the group of people who are married is University.	
Q8.	The variable "education level" is binary (dichotomous).	
Q9.	The variable "marital status" is binary (dichotomous).	
Q10.	There are more males than females in the sample.	

Let us now consider the answers:

O1 – true. We can see that the t statistic for the comparison of males and females is statistically significant, indicating a statistically significant difference between the two populations. We can read that the statistical significance value – p is .03. We can also read that the Excessive use of the Internet is a dependent variable meaning that it is the variable on which the means of the groups were calculated and with regard to which the comparison was made.

O2 – true. Similar to O1, but we should look at the p value for marital status.

O3 – false. The size of the correlation coefficient used as an effect size measure is .171 and this is considered a low correlation in all the recommendations for interpreting correlations mentioned in this book.

O4 – false. The size of the difference between means of the two groups on marital status is .159, as we can read from the table and this is a low correlation.

O5 – true. The test used here is the t test, which indeed is a parametric test.

O6 – false. Since we can see that t test was conducted and means and standard deviations calculated on the values of the excessive use of internet, it could not be a nominal variable, but must be at least on the interval level of measurement.

O7 – unknown. Although homogeneity of variances is often listed as a requirement for the calculation of the t test, there is a variant for conducting the procedure when the variances are not equal. What refers to variance refers also to the standard deviation. We have no information in the table regarding the population values of standard deviations of the two groups, nor is there any test conducted to test a hypothesis about this.

O8 – unknown. The statement phrased a bit differently, in the way that the reader is likely to encounter similar statements in scientific discussions or literature, but it states the same as O7, only for the other variable. The answer is the same. There is no data about relations of standard deviations in the populations the two groups are from.

O9 – true. These are independent samples, hence adequate for Mann-Whitney's U test and whenever t test can be used, U test can be used also.

O10 – true. We can see from the table that males have a higher mean and that the difference between means is statistically significant. Therefore, we can conclude that there likely is a difference between means of the two groups in the population and it has the direction stated in the statement. It is quite small.

P – general comments. Looking at the table, we can see that it contains the results of the Mann-Whitney U test conducted to compare groups of people who have passed a vehicle driving test and those who did not manage to pass the vehicle driving test after a traumatic brain injury. That is why variables like Amnesia, Time from injury, Coma duration and similar are in the table. We can see the U statistic in the column named U, the p value i.e. statistical significance in a separate column and likely a correlation coefficient marked with r in the far right column used as a measure of effect size. However, there is no mention in the table which exactly correlation coefficient was used, but r is a standard designation for correlation coefficients in tables like this, so we will treat it as such.

P1 – false. Age is the variable in the first row (the first row after names of statistics) and we can see that the difference between the two groups is not statistically significant. The p value is .78, which is far from any commonly used statistical significance threshold. Therefore, we accept the null hypothesis and conclude that there is no age difference between groups. Yes, these are the results of a U test, so not exactly referring to the mean, but it can be interpreted to mean that there is no difference in the central tendency of the two groups. Also there is a correlation between the two groups and age that is also not statistically significant.

P2 – true. We can see that the p value for the difference of the two groups on RTAV is very low (i.e. indicating a highly statistically significant result!) and the authors also marked the difference in bold highlighting it as statistically significant.

P3 – false. We should compare the effect size measures of the two groups (r column) and we can see that, contrary to the statement, the effect size is larger on RTAV (.44 vs. .29).

P4 – unknown. While we can infer from the designation – r, that it is some type of correlation coefficient, there is no data on which one exactly. It might be the rank-biserial, it might be Spearman correlation, but it might also be a biserial correlation – we do not know for certain from the data at hand.

P5 – unknown. While we can see that the two groups differ on this variable and that the groups are formed based on whether they have passed the test or not, we do not know whether the difference on this particular variable is due to driving skills or something else.

P6 – meaningless. There is no such thing as a "median significance score" that could be high for a variable.

P7 – false. The Mann-Whitney U test was used and it is a nonparametric test.

P8 – true. We can see that the difference between the two groups on the variable age is not statistically significant.

P9 – true. We can see that the results of the U test are statistically significant for this variable with p value being very, very low and also that the effect size is very substantial. This means that we should indeed reject the null hypothesis. People who managed to pass the driving test after a traumatic brain injury and those who did not, tend to differ quite a bit on the time they spent in coma, as results presented here show.

P10 – false. We can see that the U statistic is statistically significant for this variable, indicating that we should reject the null hypothesis, contrary to what the statement says.

Q – general comment. The table presents the results of a chi square test testing the null hypothesis that there is no correlation between variables Marital status and Education. We can see that Marital status is a binary variable, while Education is a nominal variable with 3 categories. Aside from the chi square statistic, Cramer's V correlation coefficient is also calculated and presented.

Q1 – false. We can see that both Cramer's V and chi square are statistically significant, thus indicating that the two variables for which these were calculated are associated i.e. that their association is statistically significant.

Q2 – true. According to Cohen's recommendations the correlation of .388 is indeed a moderate one.

Q3 – true. We can see that its three categories are University, Secondary school and Other. While it might be said that University is a higher level of education than Secondary school, such a relationship does not stand for the category other, hence this is a nominal variable (not an ordinal!). It is also used as such in the presented calculations.

Q4 – false. We can see that of all the married participants in the sample, only 14.3% have secondary education, but this percentage is 56.1% among participants who are not married.

Q5 – false. We can read in the bottommost row that there are 188 entities (study participants in the sample).

Q6 – false. Cramer's V was calculated, which is correlation coefficient for nominal variables, hence no meaningful negative values.

Q7 – true. Indeed, University is the most common education level category in the category of married study participants. We can see, that 83.7% of married participants reported having university education.

Q8 – false. We can see that Education in this table has 3 categories. Therefore, it is not binary as binary variables have 2 categories.

Q9 – true. We can see that there are only two categories of Marital status in the table, hence it is binary variable here. It should be noted that this statement is exclusively valid for this particular table as both education level and marital status can be conceptualized with a different number of categories than the number presented here.

Q10 – unknown. There is no data in the table on the genders of study participants.

References

Čižman Štaba, U., Klun, T. R., & Robida, R. (2021). Predicting Factors of Driving Abilities after Acquired Brain Injury through Combined Neuropsychological and Mediatester Driving Assessment. *Psihologija, 54*(2), 137–154. 10.2298/PSI200408024C

Cohen, J. (1988). *Statistical Power Analysis for the Behavioral Sciences, 2nd Edition*.

Field, A. (2009). *Discovering Statistics Using SPSS*. SAGE Publications Ltd.

Rakić-Bajić, G., & Hedrih, V. (2012). Prekomjerna upotreba interneta, zadovoljstvo životom i osobine ličnosti [Excessive Use Of The Internet, Life Satisfaction and Personality Factors]. *Suvremena Psihologija, 15*(1), 119–131.

8 Exercises – let us apply what we learned in this book!

Having passed all the contents of this book, let us now try to apply what we have covered in it all together through a series of exercises this chapter consists of. As with all the previous exercises, please refer to the start of the book for the general instruction for completing the exercises. Our suggestion is that you first read the excerpt and the statements and provide your own answer. You may write it in the Answer column, and after that look up the answers and compare your own answers with them.

Exercise R. General exercise. (Konrad, 2016)

Table 2
Correlation between compromise and aggressive conflict style of parents and adolescent

		Compromise				Aggression			
		father	a – f	mother	a – m	father	a – f	mother	a – m
Compromise	father	-	.663**	.389**	.425**	-.527**	-.446**	-.318**	-.296**
	a-f		-	.486**	.662**	-.352**	-.371**	-.303**	-.313**
	mother			-	.701**	-.222**	-.271**	-.561**	-.475**
	a-m				-	-.157**	-.215**	-.333**	-.307**
Aggression	father					-	.719**	.457**	.498**
	a-f						-	.560**	.689**
	mother							-	.788**
	a-m								-

Note. ** $p < 0.01$; a-f – adolescent – father conflict; a-m – adolescent – mother conflict;

Data were collected from 514 adolescents who responded to a questionnaire on conflict styles with their mother and father. Compromise and aggression are two conflict styles and variables father and mother added to them refer to how much the adolescent perceived them as applying that conflict style in conflicts with them, while the a-f and a-m variables refer to how much the adolescent perceives him/ herself as using that conflict style. For example, compromise a-f refers to how much adolescent perceives him/herself using a compromise conflict style in a conflict with his/her father.

Table reprinted from Konrad, S. Č. (2016). Parent-adolescent conflict style and conflict outcome: Age and gender differences 2. Psihologija, 49(3), 245–262. https://doi.org/10.2298/PSI1603245C with the permission of authors.

DOI: 10.4324/9781003107712-8

R	Statement:	Answer
R1.	The variable adolescent-father conflict is on the nominal level of measurement.	
R2.	Adolescents who use compromise conflict style in conflict with their father tend to use more the aggression conflict style in conflicts with their mother.	
R3.	Correlation between Aggression father and Aggression mother is lower than correlation between compromise mother and Aggression mother.	
R4.	Adolescents tend to report their conflict style with a parent as similar to the conflict style used by the parent.	
R5.	There were more female than male participants in this study.	
R6.	Adolescents from the sample tend to use the Compromise conflict style more often than the Aggression conflict style.	
R7.	All correlation coefficient presented here are statistically significant.	
R8.	Variables related to the Compromise conflict style are all positively correlated to the variables related to the Aggression conflict style	
R9.	Correlation coefficients below .3 are not statistically significant on this sample.	
R10.	Fathers use Aggression conflict style more often than mothers.	

Exercise S. General exercise. (Kwon & Lee, 2020)

Table 6
Paired-sample T-test.

Paired-sample (n = 252)	Time	Mean		SD		t		p-value	
		LS	A	LS	A	LS	A	LS	A
T1-T2	T1	4.51	4.67	1.47	1.49	0.565	.797	.573	.426
	T2	4.47	4.61	1.39	1.36				
T1-T3	T1	4.51	4.67	1.47	1.49	.367	-.540	.714	.590
	T3	4.49	4.71	1.34	1.33				
T1-T4	T1	4.51	4.67	1.47	1.49	3.074***	2.571**	.002	.011
	T4	4.28	4.48	1.46	1.40				
T2-T3	T2	4.47	4.61	1.39	1.36	-.243	-1.610	.808	.109
	T3	4.49	4.71	1.34	1.33				
T2-T4	T2	4.47	4.61	1.39	1.36	2.904***	2.066*	.004	.040
	T4	4.28	4.48	1.46	1.40				
T3-T4	T3	4.49	4.71	1.34	1.33	3.213***	3.573***	.001	.000
	T4	4.28	4.48	1.46	1.40				

*T1: 15th day before Travel, T2: immediately after travel, T3: 15th day after travel, T4: 30th day after travel.

Table presents data from a longitudinal study of life satisfaction before and after travelling. Variables LS and A are Life satisfaction and affect and T1-4 are timepoints when these variables were measured.

Table reprinted from: Kwon, J., & Lee, H. (2020). Why travel prolongs happiness: Longitudinal analysis using a latent growth model. Tourism Management, 76, 103944. https://doi.org/10.1016/j.tourman.2019.06.019, with the permission from Elsevier, number 5132920194933

S	Statement:	Answer
S1.	Life satisfaction tends to be higher before travel than 30 days after travel.	
S2.	Life satisfaction is higher immediately after travel than 15 days after travel.	
S3.	On the sample, the variance of life satisfaction was highest 15 days after travel.	
S4.	Wherever the difference on life satisfaction was statistically significant, so was the difference on Affect.	
S5.	The prospect of travelling improves life satisfaction.	
S6.	The test used here is a nonparametric one.	
S7.	The samples compared here are paired samples.	
S8.	In the population this sample is from, life satisfaction remains the same before and 15 days after travel, but becomes lower 30 days after travel.	
S9.	The only statistically significant differences between means are between the measurements at T3 and the remaining timepoints.	
S10.	In the population, Affect immediately before travel is lower than in all other timepoints.	

Exercise T. General exercise. (Han et al., 2018)

Table   **Pearson correlations of the creativity tasks, age, and intelligence.**

p-value	Flexibility	Originality	RAT	CAQ	Age	Intelligence
Fluency	0.63**	0.90**	0.02	0.23**	−0.14*	0.18**
Flexibility		0.47**	0.02	0.14*	−0.16**	0.11
Originality			−0.01	0.20**	−0.15*	0.18**
RAT				−0.01	−0.19*	0.19**
CAQ					−0.11	−0.01
Age						−0.07

Notes:
* p < 0.05.
** p < 0.01.

Table reprinted from: Han, W., Zhang, M., Feng, X., Gong, G., Peng, K., & Zhang, D. (2018). Genetic influences on creativity: An exploration of convergent and divergent thinking. PeerJ, 2018(7). https://doi.org/10.7717/PEERJ.5403, with the permission of authors.

T	Statement:	Answer
T1.	The Barlow's coefficient is highly significant for both CAQ and Age	
T2.	All 5 variables listed in the 5 first rows decrease with Age.	
T3.	Originality and Fluency are highly correlated.	
T4.	Intelligence is highly correlated with RAT.	
T5.	The association between Fluency and CAQ is nonmonotonic.	
T6.	RAT has no statistically significant correlations with the other variables.	
T7.	People with higher Intelligence also tend to have higher Fluency.	
T8.	Data presented here are on the ordinal level of measurement.	
T9.	The levels of Fluency in this sample are higher than levels of Originality.	
T10.	The null hypothesis tested here is that correlation coefficient in the population equals 0.	

Exercise U. General exercise. (Vukčević et al., 2016)

Table 1

Means, standard deviations and metric properties of scales

	i	M	SD	Md	Min	Max	K-S (p)	Sk	Ku	α	KMO	H1	H5
HTQ Part I	64	23.44	8.32	24	1	53	.06 (.39)	.00	.90	.89	.93	.11	.39
HTQ Part I short	48	21.40	7.60	23	0	42	.09 (.06)	-.37	.45	.87	.92	.13	.47
HTQ Part III	6	1.38	1.33	1	0	6	.24 (.00)	1.09	1.08	.56	.71	.21	.70
HTQ Part IV Total	40	93.80	22.27	92	42	149	.07 (.28)	.03	-.69	.92	.96	.22	.55
HTQ Part IV PTSD	16	40.24	9.55	40	16	61	.06 (.36)	-.20	-.57	.82	.91	.22	.57
HTQ Part IV Functioning	24	53.56	14.31	53	24	93	.08 (.16)	.21	-.63	.89	.95	.25	.61
HSCL-25 Total	25	57.67	17.06	58.8	25	96	.06 (.44)	.03	-.86	.93	.98	.34	.68
HSCL-25 Anxiety	10	22.01	7.77	22	10	40	.08 (.14)	.14	-.75	.87	.96	.41	.79
HSCL-25 Depression	15	35.76	10.36	37	15	58	.08 (.15)	-.10	-.86	.88	.96	.38	1
BDI-II	21	23.04	12.58	22.5	0	54	.06 (.39)	.18	-.78	.91	.97	.32	.66

Note. Md – Median, K-S – Kolmogorov-Smirnov D statistic, df was 226 for all variables, α – Cronbach's alpha internal consistency coefficient, KMO – Kaiser-Mayer-Olkin measure of sampling adequacy, H1 – average item intercorrelation, H5 – Knežević-Momirović homogeneity measure (Knezević & Momirović, 1996)

Table reprinted from: Vukčević, M., Momirović, J., & Purić, D. (2016). Adaptation of Harvard Trauma Questionnaire for working with refugees and asylum seekers in Serbia. Psihologija, 49(3), 277–299. https://doi.org/10.2298/PSI1603277V, with the permission from authors.

U	Statement:	Answer
U1.	HTQ part III has a platykurtic distribution.	
U2.	The distribution of all variables presented here is normal except HTQ part III	
U3.	The Chi square test was used here to test the normality of the distribution.	
U4.	Upper boundary of the 95% confidence interval of standard deviation of HTQ Part I would be higher than 30.	
U5.	Depression BDI-II and HTQ Part I are highly correlated.	
U6.	Mean of HTQ part III is higher than the median.	
U7.	The distribution of Anxiety HSCL-25 is normal in the population.	
U8.	HTQ part III has a positively asymmetric distribution.	
U9.	50th percentile of the variable HTQ Part I is lower than 30.	
U10.	HTQ Part I short has higher intermediary range than the median harmonic.	

And for the final exercise, let us try interpreting a table presenting statistics that were not covered in this book! Statistical procedures constantly evolve and authors of scientific texts will often introduce new statistical procedures and statistical measures that were not common or not used in previously published textbooks at all. But the same principles of statistics work with them also. The statistic used here is the hazard ratio, which is very similar to the odds ratio, discussed in the book. The hazard ratio is the ratio of two hazard rates. It shows how many times more likely an event is to happen to one group than to another during a certain interval. The way it is presented in this example, it represents the comparison of likelyhoods of the onset of diabetes at a certain age in groups with different listed properties. One group has a baseline value of 1, and the ratios of other groups show how much more or less often the event happens to that group. The null hypothesis p values refer to is that all compared groups have the same likelihood of the event and we can also observe whether it can be accepted by observing the confidence interval of hazard ratio of one group and see whether the value of another group (to be compared) is within that interval. Higher hazard ratios here indicate that diabetes tends to be first diagnozed at an earlier age with that group.

Exercise V. General exercise. (Noh et al., 2018)

Table 2. Adjusted hazard ratios for the association between family history and age of onset for type 2 diabetes mellitus

		HR	CI
Family history of T2D	No	1.00	
	Yes	1.37	(1.20–1.56)***
Sex	Women	1.00	
	Men	1.11	(0.93–1.33)
BMI	Underweight	1.00	
	Normal	0.98	(0.54–1.79)
	Overweight	1.03	(0.56–1.89)
	Obese	1.20	(0.65–2.20)
Exercise	No	1.00	
	Yes	1.23	(1.08–1.40)**
Smoking status	Never	1.00	
	Past	0.87	(0.72–1.05)
	Current	1.62	(1.32–1.99)***
Drinking	No	1.00	
	Yes	1.32	(1.13–1.54)**

HR: Hazard ratio; CI: Confidence interval; T2D: Type 2 diabetes mellitus.
***$p<0.001$; **$p<0.01$; *$p<0.05$; ns: not significant.

The study participants were patients receiving treatment for blood sugar levels.

Table reprinted from: Noh, J.-W., Hee Jung, J., Eun Park, J., Hwa Lee, J., Hee Sim, K., Park, J., Hee Kim, M., & Yoo, K.-B. (2018). The relationship between age of onset and risk factors including family history and life style in Korean population with type 2 diabetes mellitus. The Journal of Physical Therapy Science, 30, 201–206, with the permission from the Society of Physical Therapy Science.

V	Statement.	Answer
V1.	Patents with family history of diabetes were on average diagnosed with diabetes at the same age as those without family history of diabetes.	
V2.	In the population, men are, on average, first diagnosed with diabetes at the same age as women.	
V3.	People who have a history of drinking alcohol tend to be diagnosed with diabetes at a later age than those who do not have a history of drinking.	
V4.	Past smokers tend to have been diagnosed with diabetes at an earlier age than those who never smoked.	
V5.	Current smokers tend to have been diagnosed with diabetes at an earlier age than patients who never smoked.	

(Continued)

V	Statement:	Answer
V6.	Patients who report exercising tend to be diagnosed with diabetes at an earlier age than those who report that they do not exercise (do physical exercises).	
V7.	Patients of normal BMI (Body mass index) tend to be diagnosed with diabetes at an earlier age than patients who are underweight.	
V8.	Obesity causes diabetes.	
V9.	Drinking alcohol causes diabetes.	
V10.	Smoking causes diabetes.	

Let us now consider the answers:

R1 – false. There are negative correlations, therefore cannot be nominal.

R2 – false. Correlation between compromise a–f and Aggression a–m is negative, not positive (-.313). This indicates that those adolescents use aggression conflict style less in conflict with their mother.

R3 – false. The first correlation .457 and the other is -.561. The first is a higher number. However, when comparing correlation size we disregard the sign, as it just indicates the direction of the association and not intensity. When we do that .457 is lower than .561.

R4 – true. We can see that all correlations between a certain conflict style of an adolescent in conflict with a parent and that conflict style by that parent are positive. Those are actually the highest correlations in the table. For example, compromise father and compromise a–f is .663, compromise mother and compromise a–m is .701 etc.

R5 – unknown. There is no data on the distribution of genders in the table.

R6 – unknown. The table contains correlations, there is no data about the levels of these conflict styles.

R7 – true. They are all marked with ** and most of them are rather high.

R8 – false. We can see in the table that these correlations are negative.

R9 – false. We can see that a couple of coefficients lower than .3 are also statistically significant.

R10 – unknown. Data contains correlations, measures showing us only which variables tend to change together. We do not know their intensities. If one variable has a very low intensity and another very high, they can still be highly correlated. Correlation is not affected by intensity.

S1 – true. Mean of LS at T1 is 4.51, while it is 4.28 at T4 and the p value is .002, meaning that T1 is higher and this difference is statistically significant.

S2 – false. The difference between means of Life satisfaction – LS at T2 and T3 is not statistically significant – p value is .808.

S3 – false. Variance is the square of standard deviation and standard deviation of LS at t3 is 1.34, which is actually the lowest value of all 4 timepoints.

S4 – true. Yes, all three comparisons that resulted in statistically significant differences on LS are the same comparisons where differences on Affection were statistically significant also.

S5 – unknown. The table presents the differences in mean Life satisfaction and Affection at 4 timepoints, there is no data on causal relationship mentioned in the statement. On the sample, mean of life satisfaction 15 days before travel is the higher than at any other timepoint (though the difference is not statistically significant compared to T2 and T3), so it might have something to do with the expectation that they will be travelling.

S6 – false. This is t test, a parametric test.

S7 – true. Yes, these are the same people who underwant assessment at four timepoints.

S8 – true. The only statistically significant difference is between the mean of LS 30 days after travel i.e. at T4 and all other timepoints. We accept the null hypotheses for the other timepoints and consider that means of LS are the same at all the other timepoints.

S9 – false. The only statistically significant differences are between means of both variables at T4 and at other timepoints, not T3.

S10 – false. The results show that the mean of Affect at T1 differs statistically significantly only from T4.

T1 – meaningless. No such thing as Barlow's coefficient that could be statistically significant exists.

T2 – false. The last correlation, with CAQ, although also negative, is not statistically significant. This means we accept the null hypothesis and treat it as 0 in the population.

T3 – true. Yes, we can read that their correlation is .9.

T4 – false. Although their correlation of .19 is statistically significant, it is not high.

T5 – false. The Pearson correlation coefficient used here is a linear correlation, so unless the authors used it inadequately, the relationship must be linear (the instruction is that we assume that statistical procedures that are presented were correctly applied).

T6 – false. It has with Age and Intelligence. Although low, they are statistically significant.

T7 – true. The correlation is low, but statistically significant.

T8 – false. The Pearson correlation coefficient used here requires at least the interval level of measurement.

T9 – unknown. The data in the table are correlations, there is no data presented in the table about the level of expression (it might be elsewhere in the original article, but is not present in this excerpt).

T10 – true. That is indeed the null hypothesis when calculating statistical significance of the correlation coefficient.

U1 – false. Kurtosis is 1.08 indicating a leptokurtic distribution.

U2 – false. Results of KS are statistically significant for HTQ Part I short also (p value is .06), so we can conclude that this variable also is not normal in the population and not just the distribution of HTQ part III. Since the table does not mention which theoretical distribution was used as a reference for K-S, we assume that it was the normal distribution as this is the most commonly tested in the literature and the expected distribution of individual differences such as those represented by these variables.

U3 – false. No, K–S test was used.

U4 – false. The standard deviation is 8.32, so the standard error of standard deviation must be much, much smaller and 30 is more than 2 standard deviations above the value of the sample standard deviation (upper boundary of the 95% confidence interval is roughly 2 standard errors above the standard deviation of the sample).

U5 – unknown. There is no data in the table on the correlations between the listed variables.

U6 – true. Yes, mean is 1.38 and median is 1. And even if these were not listed, we could have concluded that from the fact that skewness is positive (indicating that mean is higher than median).

U7 – true. We can see that K-S test for this variable is not statistically significant, p value of the test is .15. So, we accept the null hypothesis that the distribution is normal in the population.

U8 – true. Skewness is positive indicating a positively asymmetric distribution.

U9 – true. 50th percentile is the median and it is 24, which is lower than 30.

U10 – meaningless. There are no such things as "intermediary range" or "median harmonic" in this context.

V1 – false. We can see that patients with a family history of diabetes type 2 (T2D) have a higher HR and that the difference is statistically significant. Value of the No group, which is 1, is also outside the confidence interval of the Yes group which is 1.20-1.56.

V2 – false. We can see that the difference between their HRs is not statistically significant, therefore we accept the null hypothesis.

V3 – false. HR for the drinking groups is higher and the difference is statistically significant. However, the hazard ratio is higher for the drinking group indicating that they tend to be diagnosed at an earlier age.

V4 – false. We can even see that their HR is lower than for those who never smoked. However, the difference is not statistically significant.

V5 – true. We can see that this difference is statistically significant and the HR of those who never smoked is outside the confidence interval for the hazard ratio of current smokers. The difference is also in the specified direction.

V6 – true. We can see that the difference is statistically significant and in the specified direction.

V7 – false. The difference is not statistically significant and their HRs on the sample are almost the same.

V8 – unknown. The data show differences on the sample with the obese group having a larger HR than the other categories, but the difference is not statistically significant in this study and we can see that statistics of all the other groups are within the confidence interval of the obese group, which itself is very wide.

V9 – unknown. Possible, but we cannot draw causal conclusions just from these data.

V10 – unknown. Also possible, but the situation is the same as with V9.

References

Han, W., Zhang, M., Feng, X., Gong, G., Peng, K., & Zhang, D. (2018). Genetic Influences on Creativity: An Exploration of Convergent and Divergent Thinking. *PeerJ, 2018*(7). 10.7717/PEERJ.5403

Konrad, S. Č. (2016). Parent-adolescent Conflict Style and Conflict Outcome: Age and Gender Differences 2. *Psihologija, 49*(3), 245–262. 10.2298/PSI1603245C

Kwon, J., & Lee, H. (2020). Why Travel Prolongs Happiness: Longitudinal Analysis Using a Latent Growth Model. *Tourism Management, 76*, 103944. 10.1016/j.tourman.2019.06.019

Noh, J.-W., Hee Jung, J., Eun Park, J., Hwa Lee, J., Hee Sim, K., Park, J., Hee Kim, M., & Yoo, K.-B. (2018). Kyung Hee University Hospital at Gangdong, Republic of Korea 6) Diabetes Education Unit. *The Journal of Physical Therapy Science, 30*, 201–206.

Vukčević, M., Momirović, J., & Purić, D. (2016). Adaptation of Harvard Trauma Questionnaire for Working with Refugees and Asylum Seekers in Serbia. *Psihologija, 49*(3), 277–299. 10.2298/PSI1603277V

Index